LOGIC, TRUTH AND THE MODALITIES

SYNTHESE LIBRARY

STUDIES IN EPISTEMOLOGY,

LOGIC, METHODOLOGY, AND PHILOSOPHY OF SCIENCE

VOLUME 278

J.N. MOHANTY

Temple University,
Philadelphia and Emory University, Atlanta

LOGIC, TRUTH AND THE MODALITIES

From a Phenomenological Perspective

KLUWER ACADEMIC PUBLISHERS

DORDRECHT / BOSTON / LONDON

A C.I.P. Catalogue record for this book is available from the Library of Congress.

ISBN 0-7923-5550-4

Published by Kluwer Academic Publishers,
P.O. Box 17, 3300 AA Dordrecht, The Netherlands.

Sold and distributed in North, Central and South America
by Kluwer Academic Publishers,
101 Philip Drive, Norwell, MA 02061, U.S.A.

In all other countries, sold and distributed
by Kluwer Academic Publishers,
P.O. Box 322, 3300 AH Dordrecht, The Netherlands.

Printed on acid-free paper

Printed in the Netherlands.

for

Günther Patzig,
Göttingen

a friend for five decades

TABLE OF CONTENTS

PREFACE

This volume is a collection of my essays on philosophy of logic from a phenomenological perspective. They deal with the four kinds of logic I have been concerned with: formal logic, transcendental logic, speculative logic and hermeneutic logic. Of these, only one, the essay on Hegel, touches upon 'speculative logic', and two, those on Heidegger and König, are concerned with hermeneutic logic. The rest have to do with Husserl and Kant. I have not tried to show that the four logics are compatible. I believe, they are--once they are given a phenomenological underpinning.

The original plan of writing an Introduction in which the issues would have to be formulated, developed and brought together, was abandoned in favor of writing an Introductory Essay on the 'origin' — in the phenomenological sense — of logic.

J.N.M.
Philadelphia

INTRODUCTION: THE ORIGIN OF LOGIC

The question of the origin of logic may pertain to historical origin (When did it all begin? Who founded the science of logic?), psychological origin (When, in the course of its mental development, does the child learn logical operations?), cultural origin (What cultural — theological, metaphysical and linguistic — conditions make such a discipline as logic possible?), or transcendental-constitutive origin (What sorts of acts and/or practices make logic possible?).

The historical question can be securely answered only within the circumscribed limits of a particular culture. Plato, more importantly Aristotle, may have laid the foundation of Greek and Western logic. But the question would remain open: did the Hindu and Buddhist logics originate independently of the Greek logic? Only the self-assured Eurocentrist can dogmatically assert that logic is, and has to be, Western in origin. The cultural question likewise can be, with confidence, answered within a given culture: e.g., the origin of logic lies in the "Logos" doctrine, the "Nous" conception, of the Greeks. What then about the Hindu and the Buddhist logics? Shall we look for a counterpart of "Logos" and "Nous" in the Orient? In the *ṛta* of the Vedic hymns, or *vâk*, meaning speech and thought? What to do with Chinese logic? Our favored genealogies begin to flounder.

But three things remain despite these frustrations: language, learning, intellectual acts and practices. Anything non-linguistic — sensations, feelings, ineffable experiences — falls outside of logic, and logic could not have arisen out of them. Linguistic sophistications of some sort or another are required for logic of any sort to be possible. Learning also is indispensable: the infant, then the child and the young adult must be able to learn, acquire and appropriate skills and practices without which the abstractions and operations on abstractions — both needed in logic — would not make sense. Linguistic achievement must keep pace with this learning. There must be, besides, fundamental intellectual acts (idealization and nominalization amongst them) but for which the abstract entities such as concepts and propositions could not be made available. And these three — developmental, linguistic and transcendental-constitutive — must be closely related to each other. It is in this relationship that the origin of logic, taken in its basic concepts, must lie. Once this infrastructure is available, technicians can build higher level structures.

Of the basic concepts whose availability is not to be taken for granted and whose "origin" is what the question of the origin of logic is all about, two seem to be most fundamental: these are "meaning" and "truth". In a certain sense of "truth", standard logic may be regarded — as it was by Frege — as explication of the formal conditions of truth. In this sense, this logic is truth-functional. But "truth" in this sense belongs not to concrete, particular speech utterances or inscriptions nor to particular, concrete entities ("a true friend"), but to meanings of utterances/inscriptions, to ideal propositions (or, if you suspect propositions, you may find, depending upon your ontological predispositions, "eternal sentences", "classes of sentences", or some such, more welcome). Once you have the concept of meaning and so propositions, and the concept of truth as it applies to them, you have almost all that is required to let logic originate. Can we say anything about how after all we come to this stage in the first place?

AN ONTOGENETIC THEORY

Following the researches of Piaget and his school, it is possible to sketch the outlines of an ontogenetic theory of logic, which would be a developmental theory of formal-logical thinking. For this purpose, I will draw upon the little book by Piaget entitled *Logic and Psychology.*[1] In order to show how children develop logical thinking, Piaget has to begin by setting aside the commonly held view that logical relationships are primarily relationships amongst linguistic expressions, or even that logic primarily concerns a system of universal, ideal entities. Originally, for Piaget, it is a group of operations that are logical. The logical relationship is the resultant of the reversible actions of composition and decomposition (e.g., the operation on two sets, A and A' which produces a new set B, or the reverse operation applied to B which splits it into A and A'), operations of grouping and regrouping carried out on objects. What Piaget does is first to construct a psychological theory of operations, then to examine logical operations, treating them as algebraic calculi and as structured wholes (i.e., systems of elements defined by a general set of laws, such as the laws which define a group or a lattice), and then to compare the two results to demonstrate their correspondence, and proposes a structuralist theory to account for this correspondence.

At present, I am concerned not with this structuralist theory, but with the developmental theory. According to this theory, there are several steps by which one arrives at logical thinking. First, through the second year of the infant, there is no thinking and so no logical thinking. At this sensori-motor phase, symbols have not been dissociated from objects. Clearly where symbols are not detached from things they symbolize, there is no thinking. At the second, i.e., the pre-operational thinking phase, extending up to five or six years of age, the first detachable symbols, which may be called proto-symbols, are constructed, but these symbols have still personalized, affectively charged, idiosyncratic meanings. This phase leads to gradually generalized meanings, but as Erikson noted, the child is capable of misunderstanding everything he hears, he has a sort of illusion of logic. The third phase, called the period of concrete operational thought, from five or six up to ten or twelve years of age, is marked by primarily verbal thinking, and follows the laws of syntax and logic, but the logic is tied to concrete facts, objects and events, embedded in the context of linear causes and effects. Adolescence brings about a phase of formal operational thinking, a general system of logic. A reflective point of view allows the adolescent to go beyond the constraints of actual objects and events. The process of emancipation continues to the world of possibilities.

Piaget insists that sensori-motor intelligence is not yet operational, for the child's actions have not yet been internalized in the form of representations. But the child is able to construct certain invariants — which shows a tendency towards reversibility. Thus, a permanent object comes to be constructed. The spatial field comes to be so organized that the child is able to return to his starting point, thereby exhibiting reversibility. A group of such displacements is thus formed — "group" in the mathematical sense. A permanent object is such an invariant group. In the second, preoperational phase, one finds symbolic play, fictional invention, deferred imitation, mental imagery, representation formation — all connected with language. But there is an absence of reversible thought operations, and absence of invariants save at the sensori-motor level. The ideas of quantitative identity, e.g., of length and transitivity (if A = B, B = C, then A = C) have not yet been arrived at. It is at the third phase that actions of combining, dissociating, ordering and setting up of correspondences become reversible operations: there are elements of logic of classes and relations, but the child does not as yet have all possible transformations of classes and relations. Each field of experience (e.g., of shape and size and weight, etc.) is given a structure by a group of concrete operations and invariants. But what is still lacking is that these operations and invariants cannot be generalized over all fields at once. We still do not have a formal logic. At the fourth level, the ability to reason ("If ... , then .... .") from hypothetically posited data shows itself and shows that thought does not proceed from the actual to the theoretical but starts from theory and moves to the actual for verification. The thinking is now concerned with propositions rather than merely objects.

What is the relation of the logical *system* (of rules and valid argument forms) to the actual psychological thinking process of the mature individual? It seems there are two main approaches to this question. One regards logical systems as models of mental processes, as representations of normative, idealized, abstract operations. Thus, Piaget at first regarded the propositional calculus itself as a description of competence, and later looked upon the natural deduction system as using inference rules rather than axioms as a better model. But in no case any such system describes a psychological reality. The other approach, which is influenced by information processing and computational models, regards logics directly as being models of psychological reality. Clearly both accounts are one-sided. The two theories have been sought to be combined by holding that competence describes the epistemic, universal subject while procedures describe the psychological, individual, subject.[2] Piaget himself has suggested that the competence system (abstract and idealized) functions to *understand* the world, the procedure system functions to assure problem solving success.[3]

HERMENEUTIC THEORIES

There are several hermeneutic theories regarding the origin of logic — they all go back to Dilthey and Heidegger. Dilthey, for example, wants to trace back the categories of logic to the categories of life. Thus, the logical concept of predication is traced back to the *Lebensbegriff* of selfsameness or *Selbigkeit.* Such concepts as "thing", "effect", and "substance" are fundamentally categories of life. The categories of "essence", "*Bedeutung*" and "*Sinn*" are abstractions from the "center of the life-structure itself". Life has a silent knowledge of itself, which Dilthey even calls "thinking". Following Dilthey, Misch distinguishes between discursive logic and hermeneutic logic — the former is concerned with the "ontic" (whatever is given in outer perception), the latter is concerned with the inner understanding of life, of goods and values, and of the meaning and significance of situations. Misch also calls the latter "practical knowledge". Hermeneutic logic is life's interpretation of itself.

The most influential version of such a grounding of logic is given by Heidegger. This is embodied in his thesis that the apophantic "*as*", the *as* of predication, is derived from, and in this sense presupposes the hermeneutic *as*.[4] This thesis again may be elaborated in several steps. First, it is maintained that a judgment's *being about* an entity is possible only because that entity has been disclosed prior to judgment. Second, predication presupposes an original unity of the entity which has been broken up and separated into elements which are then brought together. Third, judgmental truth (as adequacy) presupposes an originary sense of truth as disclosedness of entities. Fourth, the originary disclosedness of entities takes place within the context of man's practical concern with the world. Things have their original significance in what we have got to do with them; a pencil is meant for writing, a hammer for driving nails, etc. The original disclosure of entities is not to an epistemological subject, but to a practically and affectively oriented agent. The pre-logical meaning of entities is derived from such practical projects. To such an agent, a hammer is disclosed as having the significance of that by which nails are driven. This "hermeneutic" significance is the basis of the predicative judgment "This is a hammer".

Heidegger has still to tell us *how* the entity with a practical significance *becomes* an object of theoretical judgment, or the subject of predication, how the *zuhandene Womit des Zutunhabens* is transformed into "*Worüber der aufzeigenden Aussage*".

Heidegger recognizes that between these two extremes there are many intermediate phases, all of which point back to the preconceptual, practical concern with the world. The question of the origin of the theoretical attitude is, for Heidegger, the question about the conditions of the possibility for *Dasein*'s being able to exist in the manner of scientific research.[5] To say that the theoretical arises with the disappearance of practice is neither true, nor — even if true — illuminating. What accounts for this transformation is called by Heidegger "thematization."[6] Thematization objectivates; it discloses that entities are objectively determinable — thereby making possible theoretical, predicative judgments and so logic and science. Heidegger thus gives a genealogy of logic.

We know now the path and the secret motivations of Heidegger's thinking before he arrived at the position of *Sein und Zeit*. On the one hand, he was impressed by Husserl's theory of categorial intuition, and understood it to imply that our most elementary perceptions are permeated by the logical. On the other hand, he was also taken in by Emil Lask's remark about the "homelessness" of logic — meaning that logic belongs neither to the real world nor to the ideal, for everything comes under it.[7] Both together must have, through a series of transformations, led Heidegger to the distinction between the ontic and the ontological and an ontic logic (formal logic) and an ontological logic (hermeneutic logic).[8]

If formal logic presupposes ideas of meaning, reference and truth, so does hermeneutic logic but at a more fundamental, a more primitive level. There is a pointing towards, a *Verweisung*, which characterizes an equipment and which locates an equipment within a totality of equipments. A *sign*, with which logic deals, is also, like an equipment, usable. As Heidegger writes:

> A sign is not a thing which stands to another thing merely in the relationship of indicating; it is rather an item of equipment which explicitly raises a totality of equipments into our circumspection.[9]

The meaning of a sign lies in its use, or rather usability. Predication is one of the uses in one limited context. The original context of equipmental totality has been ruptured, the original "circumspection" leveled off into an indication of entities as merely present-at-hand. The original *zuhanden* thing has become an object. As formal logic advances, it more and more gets rid of all reference to the world and becomes a logic of ideal meanings.[10]

Although Heidegger's thinking contains suggestions of a hermeneutic logic, he himself has not worked out what such a logic will look like. Following suggestions from both Heidegger and Husserl, several philosophers have worked out, in more detail, sketches of such a logic. I can only mention here Hans Lipps (who goes back to the roles that entities and structures arising out of human life play in scientific linguistic situations of everyday life, e.g., in practical life one infers not from premises but from situations), Josef König (who makes the distinction between *practical* and *theoretical* sentences as the basis of his philosophy of logic) and Paul Lorenzen (who develops a systematic constructive procedure for building up formal-logical concepts and operations from simple practical situations [such as one in which one gives an order and the other either obeys or does not; or dialogical situations in which there is a proponent and an opponent]).

There is a certain parallelism between the genetic-developmental account stated above and this hermeneutic-genealogical account: the genetic primacy of the sensori-motor activity corroborates the primacy of the practical; an original logic of dealing with things, in both cases, precedes a logic of signs as detached from things, and in both formal logic is a late bloomer. But this remarkable similarity at the same time calls into question the significance of Heidegger's thesis of the primacy of the practical.

A PHENOMENOLOGICAL GENEALOGY: HUSSERL AND GURWITSCH

Husserl, throughout his life, made several moves towards a genealogy of logic. One of his standing concerns — going back to the nineties of the last century — was: how do scientific, mathematical and logical entities arise out of the pre-scientific, pre-logical, perceptual life. The answer that famously is found in the *Crisis*[11] of his old age is to be found in some version or another throughout his *opus*: scientific objectivities are idealizations of the lifeworld. But this general answer is mediated by several moves. First, with the definitive rejection of psychologism as a theory of logic, it was possible to restore the logical to its purity as an ideal entity, now secure from all empiricistic and relativistic reduction. But this is only the beginning: the assurance and the demand that any viable genealogy must be subject to one test, namely, that it must be compatible with that ideality of the logical. The next step is to tie the thesis of ideality to the thesis of intentionality: the logical entities — coming under the general category of "meaning" — are but contents, noemata, of intentional acts. The third step, first taken in the *Formale und transzendentale Logik*, shows how complex logical entities are step-by-step constructed in intentional acts and how, through change of attitude from the direct object-directed intentions to the reflective *intentio obliqua*, meanings come to be posited as entities; how through all these constitutions, reference to the world persists inasmuch as the variables, nominative or predicative, pick up appropriate entities from the world, and how the laws of material meaningfulness and material nonsense are grounded in a transcendental aesthetic, i.e., in the structure of the lived world; and, finally, how there are idealizing presuppositions underlying all logical and mathematical propositions (as, e.g., "It can be repeated by *anyone*", "*and so on*", i.e., continuable without limit, etc.). The *fourth* step, undertaken in *Experience and Judgment*, consists in showing that the concepts and truth functions of apophantic logic have their origin in the proto-logic characterizing pre-predicative experience, that such forms as propositional negation and disjunction have their corresponding proto-logical origins. The domain of logic is thereby extended to cover pre-predicative experience where the roots of higher order logic lie.

Theoretically, a most difficult question is, how logical thinking could arise from perceptual experience. A preliminary phenomenological account would be as follows. Perceptual experience is not a chaotic succession of sense impressions, but is rather characterized by somewhat inexact typicalities. The question then is, how the transition from *type* to *concept* and *eidos* takes place. Here I turn to the researches of Aron Gurwitsch which follow Husserl's. Specific idealizing operations such as formalization and generalization, and mathematization and algebraization are to be applied in order that logic in the fully developed sense can be constituted. I will here refer to two of Gurwitsch's results: one concerns the relation between the perceptual *Gestalt* and the logical *eidos*; the other concerns the relation between perceptual apprehension (Husserl's *schlichte Erfassung*) of an object and predicative judgment. Gurwitsch, in his attempt to appropriate Gestalt psychology (especially of the Berlin school) into phenomenology, realized the gap that separated the *Gestalt* from the ideal meaning. Early on Gurwitsch distinguished, e.g., between two kinds of categorial terms (like unity and identity), one which refers to experienced data and the other conceptual and categorial. He came to recognise that the categorial is a limit to Gestalt theory.[12] But later, around 1936, he found in the researches of Gelb and Goldstein results that were helpful for understanding the relation between the precategorial and the categorial. However many predelineated and pretraced structures there may be in the perceptual domain, structures which *anticipate* the categorial structures, there must be a *leap* involved — there is no smooth natural continuity of transition.

The same happens in the transition from simple perceptual apprehension to predicative judgments. Husserl's account of this is, in brief, that from a simple apprehension of an "indeterminate substratum", there is a transition to an explicating contemplation (*explizierende Betrachtung*) in which the original theme is "maintained in grasp", but a thematizing glance is directed to one of its properties, such that there is a coincidence or overlapping between the two thematizing activities. The explicated property now accrues to the original theme as a new "layer of sense". We can return to this achievement ever again. The originally encountered thing has now acquired the sense of being a logical subject; a spontaneous act now binds them together.

Whereas Husserl understood predicative judgment as an act of synthesis, Gurwitsch views it as an act of analysis of the original object (understood as a system of noemata), while explication analyzes the system into its constituents which figure as predicates. This activity is called by him "thematization", predicative judging becomes "articulating thematization". Thematizing is originally receptive, then reflective as expressed in the copula "is". This is a categorial act and gives rise to ideal entities which Husserl, following Stumpf, had called "syntactically formed affair komplexes". We are in the domain of logic. It must be noted that here too there is a "leap" which any genealogy tends to gloss over or overlook.

SUMMING UP

We cannot speak of a causal origin of logic. It is not even clear what that might mean. One can speak of brain-states or neural information processing states giving rise to logic. Whatever that might mean, such hypotheses make sense just because the very idea of information processing neural states cannot be understood save as realizations of some logical model or other. The idea of a *logical* origin of logic begs the issue. With these two locutions out of the way, we can turn to the three major types of theories presented here.

There is first a remarkable affinity between Piaget's theory that sensori-motor intelligence precedes the concrete operational thinking and Heidegger's thesis that apophantic-predicative logic is preceded by a hermeneutic-practical understanding. However, this affinity should be considered in the light of a fundamental difference. On Piaget's theory, sensori-motor intelligence and concrete operational thinking are over by the age of twelve, while if Heidegger is right, hermeneutic-practical understanding should characterize all authentic being-in-the-world, which logical thinking degrades into a derived, inauthentic relation to things. It would be far from Piaget's intention to claim that the child's way of relating to the world is the most authentic; it would perhaps be far from Heidegger's intention to dignify the child's pre-logical understanding as the authentic being-in-the-world, although some of Heidegger's romantic views regarding "back to the origins" may suggest something like that.

Husserl also wants to return to the origin in the proto-logic of pre-predicative experience beginning with simple apprehension. But this pre-predicative experience is far from being the practical having-to-do-with tools.

There is an extreme one-sidedness in Heidegger's account of the authentic, *ursprüngliche* being-in-the-world which, precisely for that reason, has impressed and misled many: this derived from the centrality assigned to the tool, and the tool-using circumspection, and the world as a totality of equipmental references. While this is true of the carpenter's workshop or the farmer's fields, this is not all. The farmer and the carpenter also deliberate, think, assert predicative judgments, use "copulae", doubt, deny, infer (not merely from situations, but from what was said).

What I am drawing attention to is that you cannot accord primacy to the practical and derive the theoretical from it. To say that the theoretical is constituted by the act of "thematization" is not very illuminating, for thematization, on one understanding, is a theoretical act, on another it can be "practical thematization" as well. More promising would be to go beyond and behind the distinction between the theoretical and the practical, and find the roots of theoretical as well as practical idealities — of the logic of predication as well as the logic of action.

This place beyond the theoretical-practical distinction is the phenomenon of intentionality which exhibits itself as much in the proto-logic of pre-predicative experience as in the practical understanding and meaning of the tool-using carpenter or the farmer.

The origin of logic has to be sought in the structure of intentionality — *proximately*, in the structure of the living intentions of the logician, remotely in the intentions implicated in those intentionalities down to the most basic ones. I think the account would run along the following lines:

Intentional acts have ideal contents.

In the case of signitive, symbolic, intentional acts, these contents are meanings in the narrow sense.

Originally positing, these acts may be reflected upon — making explicit propositional contents.

Propositional contents must be capable of being held fast, repeatable, and retrievable.

Acts of inferring, occurring in everyday life and dialogue have complex propositional contents.

Propositional acts lead back to nominal acts having *terms* as contents.

In this way an account can be built up of how logic is constructed.

NOTES

[1] Manchester: Manchester University Press, 1953.

[2] Overton, W.F., "Scientific Methodologies and the Competence-Moderator-Performance Issue" in: E. Neimark, R. DeLisi, & J. Newman (eds), Moderators of Competence, Hillsdale, N.J.: Lawrence Erlbaum Associates, 1985.

[3] Piaget, J. Possibility and Necessity, vols 1 & 2, Minneapolis, MN: University of Minnesota Press, 1987.

[4] Heidegger, M., Sein und Zeit, Tübingen: Max Niemeyer, Seventh Edition, 1953, p.158.

[5] Ibid, p.357.

[6] Ibid, p.363.

[7] Compare Husserl's idea of formal ontology.

[8] See Theodore Kisiel,The Genesis of Heidegger's *Being and Time*, Berkeley & L.A.: University of California Press, 1993.

[9] Sein und Zeit,p.80.

[10] A nice account of this genealogy is given by Robert Hanna in his "On the Sublimity of Logic: A Heideggerian Analysis" (The Monist, vol. 69, 1986, pp. 264-280).

[11] Husserl, E., The Crisis of European Sciences and Transcendental Phenomenology, tr. D. Carr, Evanston: Northwestern University Press, 1970, p. 50.

[12] Gurwitsch, A., Studies in Phenomenology and Psychology, Evanston: Northwestern University Press, 1966.

1. THE CONCEPT OF 'PSYCHOLOGISM' IN FREGE AND HUSSERL*

Both Frege and Husserl are well-known to have campaigned against a philosophical position known as psychologism. In this essay, I will begin by focusing upon the nature of psychologism in general, and then on the specific sort or sorts of psychologism Frege and Husserl were up against, before looking at the arguments they pressed against that position. Finally, I will ask if a radical anti-psychologistic position is a defensible one.

PSYCHOLOGISM: AN ALLY OF IDEALISM OR OF NATURALISM?

It is time to recognise that Frege's attack on psychologism could not have been motivated by the idea of attacking (Hegelian) idealism. Not only had Hegel's idealism ceased to be a dominant philosophical theory by the later half of the nineteenth century, there are also examples of philosophers who regarded psychologism as their way of correcting the excesses of Hegelian idealism. Frank Kirkland cites Jakob Fries whose psychologism was advanced as an anti-Hegelian position.[1] Kirkland also cites two passages from Hegel which clearly show Hegel's extreme hostility and negative attitude towards any intrusion of psychological considerations into logic. Not unlike Frege and Husserl, Hegel insisted upon keeping *Vorstellung* and *Gedanke* apart. *Vorstellung*, for Hegel, is pictorial thinking which should be kept apart from pure thought. Pictorial thinking is an ally of naturalism, depending for its categories on things outside of logic. Pure thought derives its categories from its own internal development. Kirkland is therefore led to assert that Hegel's critique of psychologism is an instance of his critique of pictorial thinking, which is indirectly a critique of naturalism. It would then seem to be more in order of things that Frege's critique of psychologism, instead of being a critique of Hegelian idealism, was rather intended to be a part of his overall rejection of naturalism. Confusion about this link between psychologism and naturalism is made muddier by Frege's accusing Husserl, in his review of the

* Originally appeared in: *Philosophy and Rhetoric*, 30, 1997, 271-290.

latter's *Philosophie der Arithmetik*[2], of reducing everything objective to subjective, so that psychologism is maintained as leading to a sort of phenomenalism. Michael Dummett's linking psychologism to idealism--which he now thinks was a mistake[3]--was probably encouraged, not alone by Frege's misleading review of Husserl's work, but also by various other statements of Frege. One such statement is this:

> Psychological treatments of logic...lead(s) then necessarily to epistemological idealism. Since all knowledge is judgmental, every bridge to the objective is now broken off. This *flowing into* idealism is most remarkable in physiological psychology, because it is in such sharp contrast to its realistic starting point.[4]

What Frege asserts here is that psychologism leads not to phenomenalism or subjective idealism, but to "epistemological idealism," the thesis namely that knowledge is confined to ideas and cannot reach the real objects outside of the mind, and so leads to a sort of scepticism. What was more important still for Frege is that psychologism cannot preserve the objectivity of logical and mathematical cognitions for, as he wrote,

> Neither logic nor mathematics has the task of investigating minds and the contents of consciousness whose bearer is an individual person.[5]

I will not, on this occasion, attempt to decide whether Frege was a realist as Dummett takes him to be, or an objective idealist as Hans Sluga does. But one thing is clear: his anti-psychologism is motivated by the goal of preserving the objectivity of logical and mathematical truths and of setting aside the possibility of scepticism with regard to any cognition. The sort of psychologism which threatens to have these ruinous consequences is, for him, based on a naturalistic philosophy. Physiological psychology and empiricistic psychology of the British empiricists are two such theories. Neither the Kantian nor the Hegelian idealism has those consequences--not the Hegelian in particular which sought vehemently to keep psychological contents entirely out of the categories of logic.

TYPES OF PSYCHOLOGISM

Many authors have attempted a taxonomy of the various types of psychologism.[6] Without recalling those attempts, I want to begin by distinguishing 'naturalistic psychologism' from 'transcendental psychologism.' While Husserl generally used 'psychologism' for the former, he often, especially in the *Formal and Transcendental Logic*,[7] used the latter designation to stand for a variety of psychologism which his refutation in the *Prolegomena*[8] had left unaffected. It seems that by 'transcendental psychologism' he has in mind the Kantian sort of thesis that the pure forms of logic have their origin in the a priori structure of the human mind (so that pure transcendental logic has its foundation in a transcendental psychology). I will return to this variety of psychologism later in this essay, but for the present focus upon the former, i.e., naturalistic psychologism.

The "naturalism" which underlies naturalistic psychologism is, according to Husserl, a metaphysical thesis which regards the world as consisting only of individual things, cognizable by outer and/or inner perception, and the psychology based upon such a theory as one which undertakes the task of enquiring into causal relations amongst such individual mental contents and other mental contents and physical entities. This naturalism can be held by realists and idealists alike.[9] Naturalistic psychologism may then be upheld either by phenomenalists or by psycho-physical researchers or by many other asserted varieties of psychological theories which are founded upon a naturalistic world-view.[10] Both Frege and Husserl reject them all--not psychological theories *per se*, but insofar as these psychological theories claim to provide the foundation for logic and mathematics, and--only in Husserl's case--insofar as they advance a theory of the nature of consciousness.

Under the general rubric of 'naturalistic psychologism,' I would like to distinguish between various types, depending upon to which fundamental concept the naturalistic psychology is being applied. Thus we have:

A. logical psychologism, i.e., psychologism with regard to logic;

B. psychologism in theory of meaning;

C. psychologism in theory of truth;

D. epistemological psychologism; and

E. psychologism applied to metaphysics.

In each case there is a corresponding anti-psychologism, and possibly a strong anti-psychologism and a weak one.

Frege was opposed to A through C, he possibly thought he was opposed to D but as a matter of fact he was not. Husserl was opposed to A through D. They both had nothing to do with E. In this essay I will not be concerned with E which includes any attempt to use a psychological category for understanding/interpreting the nature of reality. A good example of this is Schopenhauer's characterizing Reality, the Kantian *Ding an sich*, as Will. Setting aside such attempts to ground metaphysics upon a psychological theory, let me briefly explain A through D.

Logical psychologism is the theory that logic is based upon psychology. As has been noticed by many, there are both a weak and a strong form of logical psychologism. Of the various accounts available, of this distinction, I will mention here that given by Hanna:[11] according to strong logical psychologism, all pure logical propositions (truths, principles, axioms) are necessarily dependent on the human mind, are necessarily contained in the human mind, so that logical objectivities are mental entities. On the formulation proposed by Hanna, the thesis may be stated perspicuously as:

Necessarily(x) (x is a logical proposition ---->(∃y) (y is a human being & y thinks x)) (1)

However, note that 'y thinks x' (in the above) itself allows of a psychologistic reading, meaning that if y thinks x, then x is a mental entity. If one rejects this reading and allows that 'y thinks x' does not necessarily entail that x is a mental entity, then (1) does not represent strong logical psychologism. For even if (1) is true, it may be that x is not a mental entity. So I would revise (1) into:

Necessarily(x) (x is a logical proposition ---->(∃y) (y is a human being & x is a content of y's mind)) (1*)

Weak logical psychologism, in Hanna's formulation, amounts to:

(x) (x is a logical proposition ---->M(∃y) (y is a rational thinker and y thinks x) (2)

Replacing 'y is a human being' in (1*) by 'y is a rational thinker' in (2) has the advantage that while (1*) implies anthropologism and relativism with regard to a natural kind, (2) does not. It rather uses a "structural" concept of a "rational thinker."

Now of Frege we can say that he rejects both (1) and (1*) as well as (2). Husserl certainly rejects (1), perhaps also (1*), but it may be argued, not (2).

We have noticed that 'y thinks x' allows of two readings: one psychologistic, the other not. On one reading, it means 'x is a content of y's mind,' on the other, it means 'x is an object of y's thinking.' One may want now to point out that this concerns 'meaning-theoretic psychologism.' To hold that 'y thinks x' = 'x is a content of y's mind' is to opt for a psychologistic theory with regard to meanings or thoughts, whereas to hold that 'y thinks x' = 'x is an object of y's thinking' is to opt for a non-psychologistic theory of meaning. Both Frege and Husserl rejected identification of meanings with images or any other mental entities, for both *Sinne* are objective entities which transcend the mental experiences (or acts) of any thinker.

From psychologism with regard to meanings, one can distinguish psychologism with regard to 'truth.' Both Frege and Husserl want to keep the idea of truth secure from intrusions of psychology: Frege because logic for him is an elaboration of the idea of truth, Husserl inasmuch as for him psychologism leads to relativism regarding 'truth' and so is incoherent. Psychologism with regard to 'truth' leads, according to Frege, to identifying 'true' with 'taken to be true' which he rejects but is not sure if he can refute. Husserl, on the other hand, takes enormous pains to refute the position, but concedes that he would not be able to persuade the psychologistic and relativistic thinkers.[12]

The ultimate constituents of 'truth,' for Husserl, are 'meaning' and 'object.' Meanings are, on his (and Frege's) theory, ideal entities transcending the subjective life of persons. Objects, by definition, are ontological. The adequacy of meanings to objects is truth. Truth by itself is an ideal entity, not an individual being, thing, event or fact. Even if there are no intelligent creatures to be aware of truth, truth remains in itself what it is, it retains its ideal being. It is "a unity of validity" in the timeless realm of ideas. As Frege would have said, truth is independent of human judgment. To relativise truth is to relativise being, for 'truth' and 'being' are correlative categories.[13]

Can there be an epistemological psychologism as distinguished from all the above three? If as Kitcher has maintained, Frege's epistemological theory is psychologistic,[14] and if, as would be uncontroversial, his theses on logic, meaning, and truth are anti-psychologistic, then it would seem one can distinguish between epistemological psychologism and the three others. Kitcher recognises that Frege's anti-psychologism is restricted to: preserving the objectivity of truth, guarding against identifying meanings with subjective ideas as well as against taking mathematical objects as ideas, and also rejecting the view that logic is concerned with the actual process of thinking. However, according Kitcher, all these anti-psychologistic doctrines are compatible with an epistemological theory whose basic framework was taken over from Kant's inventory of sources of knowledge and types of cognition (analytic-synthetic, a priori-a posteriori). To the extent, Kantian epistemology does not presuppose a naturalistic world view, I would add that the resulting psychologism of Frege's epistemology is not "naturalistic" in the sense we are talking about it now.

Husserl also wanted to avoid a naturalistic epistemology. Levinas expresses the basic idea of a naturalistic epistemology by saying that it "places subject and object in the same world, which it calls nature, and studies their relation as a relation of causality."[15] Husserl had to radically reformulate the problem of knowledge--in this sense, Frege's was closer to Kant's while Husserl, though Kantian in a very general sense, rejected the Kantian idea of "faculties" and "sources" of cognition and the Kantian taxonomy of judgments. The ideas of intentionality, meaning fulfillment, evidence and constitution shape both his epistemological problem and its solution--still leaving a flavor of psychologism though of a non-naturalistic sort.

Both Frege and Husserl were aware of the epistemological discussion within the Neo-Kantian circle, esp. of Marburg. Hermann Cohen had rejected psychologism in favor of an objective theory of knowledge. Cohen's program was to ground the fact of science objectively. Natorp, in a well known essay, "Über objektive und subjektive Begründung der Erkenntnis"[16] showed the superiority of objective grounding of the sciences over subjective grounding, the object being what is valid independently of the subjective. The object, for Natorp, equals the Law, which again refers to the process of scientific objectification. The scientific object is constituted inasmuch as appearances are governed by laws. The ultimate foundation of science lies in the *Gesetzlichkeit* itself. The so-called subjective foundation lacks "fundamental scientificity," lacks "insight into principles." While Natorp had welcomed Husserl's purely objective account of logic in the *Prolegomena*, Husserl's further attempts to give a phenomenological foundation for logic were criticised by him as amounting to a reversal to psychologism.

It is important to ask, in what way these forms of psychologism or anti-psychologism are mutually interconnected or interdependent. For Husserl they are all of a piece. Logic deals with ideal meanings of words and sentences, and insofar as logic is concerned with truth, truth too must be objective. Rejecting psychologism in these three areas of problems, one can possibly still subscribe to a sort of psychologism in epistemology as one does when one formulates the epistemological problem in terms of justification of beliefs.

THE ANTI-PSYCHOLOGISTIC ARGUMENTS OF HUSSERL: A RESUME

Husserl gives the most wide-ranging and painstaking critique of psychologism that is to be found in the philosophical literature. The bulk of this critique is in the chapters 4 through 8 of the *Prolegomena*, but there are also criticisms elsewhere in *Logische Untersuchungen*. For a criticism of a psychologistic theory of meaning, one should turn to Investigation 1 of *Untersuchungen*,[17] where Husserl argues for the distinction between psychological contents (images, accompanying intuitions and/or feelings) of experience, of understanding, or of meaningfully using expressions, or of thinking, and their logical contents; also between acts of meaning and the meaning intended, between judging and the proposition asserted or denied, between act of inferring and the syllogism, between having an idea and the concept entertained. Terms such as 'idea,' 'judgment,' 'affirmation,' 'denial,' 'inference' are said to be psychological or "psychologically slanted," and as playing no role in any pure theory. The scientist asserts propositions, but he says the propositions are true or false: "his premises are propositions and so are his conclusions."[18] To sum up: All theoretical science consists, in its objective content, of one homogeneous stuff: it is an ideal fabric of meanings," and meanings are "ideal unities of validity" like the theories they constitute. We can then say, if psychology itself is to be a theory, it must consist of such ideal meanings (concepts and propositions) and must not contain any "psychological stuff."

For Husserl's criticism of a psychologistic theory of truth one should refer to *Prolegomena*, sections 24, 36, 39, and 50. At all these places, while refuting relativism, individual or specific, which Husserl takes psychologism to imply, and also while refuting what he calls "anthropologism" (of Sigwart and Erdmann), he makes the same point: relativism cannot be coherent. I will return to this charge of incoherence later in this essay. For the present I would lay stress on the claim that "no truth is a fact, i.e., something determined as to time."[19] Husserl is not denying that there are changes, so he adds: "A truth can indeed have as its meaning that something is, that a state exists, that a change is going on, etc. The truth itself is, however, raised above time, i.e., it makes no sense to attribute temporal being to it, nor to say that it arises or perishes."[20] If it is true that it is raining in Ambler at time t_n this truth itself does not occur in time, the truth is timeless. The timelessness of truth would render, Husserl thinks, any relativistic, and so psychologistic theory of truth absurd. But why this should be so, we are not told by him.

Once meanings and truth are shown to be ideal entities, and so not real things (individuated by time, and by the mental life of a person) it becomes, for Husserl, easier to reject psychologism with regard to logic. The criticism also shows that already in *Prolegomena* Husserl was searching for a middle ground between psychologism and anti-psychologism, while conceding that the truth lies more on the anti-psychologistic side.[21] Chapter Three which reviews a series of arguments and counter-arguments by the opposed parties brings out that the anti-psychologistic logicians were misled when they thought that they could refute psychologism by placing emphasis on the normative conception of logic (which, by the way, is one of Frege's arguments against psychologism).[22] As a matter of fact, it might seem as though Husserl's sympathy lies with the psychologistic logicians' response that every "ought" rests on an "is." But the psychologistic logician, taking logic as a technic, a *Kunst*, argues that no technology can ignore causal connections. Husserl, however, uses "ought presupposes is" to maintain that every normative discipline presupposes a theoretical discipline. The normative sentence "An A ought to be B" presupposes the theoretical sentence "An A that is not B is a bad A." Normative logic, in this sense, presupposes "pure logic." From Frege's point of view, it is not necessary to distinguish between the ideal and the normative. Husserl would not only distinguish between them, but by grounding the normative upon the ideal more successfully overcomes psychologism.

Chapter Four of *Prolegomena* draw out the consequences of psychologism. Psychology being an empirical science, its laws cannot be exact, cannot help being vitiated by vagueness. Logical laws, on the other hand, are exact. Psychologism in theory of logic would have the absurd consequence of rendering the laws of logic vague and inexact. Secondly, no natural law can be known a priori, they are all established by induction. Logical laws, on the contrary, are a priori established by "apodictic evidence" (an idea to which I will return later in this essay). This distinction would be obliterated by psychologism. Third, psychologism construes logical laws such as the well-known principle of non-contradiction as natural laws governing human thinking in a rather causal manner. But one cannot establish a causal law by apodictic evidence. Moreover, in order to exercise a causal influence on thinking, the law must be a judgment or a cognition, itself a mental entity, but such an assumption already confuses between the law and one's thought, or cognition of the law. What is still more pertinent, one would be confusing between the law as a term in causation with the law as the rule of causation. To bring out this latter confusion, Husserl imagines an ideal person all of whose thinking proceeds "as logical laws require," or even a computer whose procedure would be causally explained by appealing to the initial arrangement regulated by natural laws which agree with those of arithmetic. But neither the machine nor the ideal thinker thinks in accordance with the laws. Finally, no logical law presupposes the existence of persons with mental experiences such as judging. The laws are about truths, but truths are not facts.[23]

In Chapters Five and Six, Husserl considers some specific attempts--esp. by Mill and Sigwart--to give psychologistic renderings of principles of logic, of the principle of non-contradiction and the principles of syllogism in particular, and argues why these renderings should be rejected. If the principle of non-contradiction, for example, is taken to be the empirical, psychological law that it is impossible for us to believe in an explicit contradiction, Husserl wants to know what is meant by 'impossible' in this statement, and also what are the circumstances under which opposed beliefs cannot co-exist. One would perhaps say something like "in a single consciousness," or "in a normal consciousness," or "in a reasonable person's mind," each of which is too vague to enter into the formulations of a law.Basically what the psychologistic rendering does is to replace "not-both-being-true of contradictory propositions" by "incapacity to entertain contradictory beliefs."

Chapter Seven interprets psychologism as leading to relativism, distinguishes between two kinds of relativism, individual and specific, and undertakes to refute each. Specific relativism becomes anthropologism in case one restricts the species to the human kind. Regarding individual relativism, Husserl recognises that one cannot refute it, for "refutation presupposes the leverage of certain self-evident, universally valid convictions," and yet proceeds to argue that the content of the relativist's assertion (that 'true' is always 'true for me') contradicts "what is part of the sense or content of every assertion" (namely, that if true, it is true for every one).[24] Is this then intended to be a refutation? What if the individual relativist does not accept the absolute validity of the principle of contradiction, or the claim that every assertion carries the claim that if true it is true for every one? What if the relativist of this brand does not recognise the concept of a *proposition* at all?

As against specific relativism, Husserl argues--as before--that the thesis is in conflict with its own sense, meaning thereby that 'truth for this, e.g., the human species' is an absurd idea, and the relativist gives the words 'true' and 'false' new meanings. But this argument takes for granted precisely what the relativist denies, namely, that 'truth' means 'absolute truth.' The anthropologist's thesis "If there were no such constitution (viz. the human), there would be no 'truth'" is regarded by Husserl to be absurd inasmuch as the consequence 'There is no truth' is equivalent to 'There is a truth that there is no truth' while the antecedent is the negation of a fact and so is false, not absurd. So the thesis deduces an absurdity from a meaningful antecedent.

Pierre Adler has distinguished between two kinds of criticism Husserl advances against psychologism-cum-relativism.[25] One of them is what Adler calls "the ontological critique" according to which psychologism confuses between the real laws and ideal laws. The other he calls "linguistic critique," which may also be called "pragmatic critique," for the incoherence Husserl brings out is not logical but "linguistic-pragmatic." I think Husserl has two varieties of criticisms under this heading, one of which is logical ('Logic being a theory of all theories cannot presuppose another theory viz. psychology') and the other pragmatic ('The relativist cannot assert his thesis').

One thing seems to be clear from the above exposition. Husserl's critique of logical psychologism, even if its strength is undeniable, does make use of an ontological dualism between the real and the ideal which Husserl has not yet, in *Prolegomena*, established. That is why to a careful reader he gives the impression of begging the issue, i.e., of making use in the premisses the very distinction between psychology and logic that he wants to prove. It is in the First Logical Investigation that he shows why meanings have to be ideal entities, and later in the third and fourth Investigations he shows why logical laws have to be analytic (and in what precise sense). I think Husserl's strongest arguments are those that point out an incoherence in the psychologistic and relativistic positions. The arguments are still valid against a psychologism which does not make use of the empiricistic psychology of the sort Husserl (and Frege) had in mind. Husserl's point would still be valid that if mathematics is fundamentally based upon some logic, any psychology which, like the modern cognitive science, makes use of mathematics cannot provide the foundation of logic. It must nevertheless be conceded that in *Logical Investigations*, Husserl has not refuted all forms of relativism, that he has refuted only those that are consequences of the sort of psychologism he was concerned with.

RESPONSES TO HUSSERL'S ANTI-PSYCHOLOGISM

Before briefly recalling the responses to Husserl's anti-psychologism,[26] it is important to note that the distinction between logic and psychology was first made not by Frege or Husserl, but already by Kant. Thus Kant writes: "die reine Logik...keine empirische Prinzipien" hat und "nichts...aus der Psychologie" schöpft."[27] He also writes:

> Nähmen wir die Prinzipien aus der Psychologie, d.h. aus der Beobachtungen über unseren Verstand...dies wurde also zur Erkenntnis bloss zufälliger Gesetze führen. In der Logik ist aber die Frage nicht nach zufälligen, sondern nach notwendigen Regeln; nicht, wie wir denken, sondern, wie wir denken sollen.[28]

Here Kant takes recourse to a normative conception of logic. We have referred earlier to Hegel's rejection of psychologism. Herbert and Lotze are the two major philosophers after Hegel, who also rejected psychologism. Thus writes Herbert:

> In der Logik ist es notwendig, alles psychologisches zu ignorieren, weil hier lediglich diejenigen Formen der möglichen Verknüpfung des Gedachten sollen nachgewiesen werden, welche das Gedachte selbst nach seiner Beschaffenheit zulässt.[29]

As regards Lotze, it is well known that Husserl acknowledges Lotze's influence on him in as far as his overcoming of psychologism is concerned.[30]

Amongst Husserl's contemporaries--leaving aside those who greeted the *Prolegomena* with unqualified enthusiasm--the anti-psychologism was received with mixed feelings. There were several kinds of criticisms. Some found his conception of psychology defective, others criticised his conception of logic. Some greeted his anti-psychologism, but wanted that after logic and psychology had been separated steps should be taken to bring them into a satisfying connection. Others found his arguments against psychologism defective, still others lamented Husserl's relapse into psychologism. There were also philosophers who were criticised by Husserl as being psychologistic, but who rejected this characterization.

To this last group belonged such men as Brentano and Schuppe. Husserl of course resisted offending his teacher by characterizing his position as psychologistic in the pejorative sense, but the fact that Brentano had to defend himself against such a charge shows that Brentano suspected Husserl's secret intention on this matter.[31] Schuppe had already said in 1879 that the task of logic was to separate the pure *Gedankenelemente* from not only their linguistic garbs but also from acts of thinking.[32] So he too rejected Husserl's characterisation of his position as being psychologistic. Benno Erdmann referring to Husserl's criticism of his view conceded that some of the criticisms were justified inasmuch as he himself did not correctly state his views and had distinguished logic from psychology in a rather unsatisfactory manner. Nevertheless, he added, Husserl had totally misunderstood him, so that without himself deciding who was right in this case, he would rather leave it to the "Unbeteiligten der jungen Generation."[33]

The most negative response came from Wundt.[34] Wundt begins by conceding that he did not know who coined the name "psychologism," but regarded the name "logicism" as being still more recent. Wundt then traces psychologism to British empiricism, while regarding logicism as being as old as speculative metaphysics. Husserl's program, he continues, is to expel psychologism not only from logic but also from psychology,[35] for what Husserl eventually aims at is a rationalistic psychology. He then goes on to criticise Husserl's concept of mental phenomena as acts. He finds this conception of act to be completely formal, empty of contents, and accuses Husserl of using "scientifically unexamined concepts of vulgar psychology."[36] Like his "descriptive psychology," Husserl's pure logic too is scholastic and concerned with word meanings. Husserl's so-called ideal meanings are his creations. There is no such invariant meaning. The only complement he pays to Husserl, his erstwhile student, is that he is an excellent critic such that "In Zerstören sucht er seinesgleichen."[37] But he hardly substantiates his positive thesis excepting by appealing to evidence but evidence is itself an *Erlebnis* and so a psychological phenomenon.

Sigwart whom Husserl criticised as being psychologistic defends himself against Husserl's criticisms.[38] As against Husserl's criticism that according to Sigwart 'truth' requires an intelligence which thinks it true, Sigwart contends that 'true' or 'false,' in the original sense of those words, can be ascribed only to an assertion or an opinion and that an assertion or an opinion necessarily presupposes a thinking mind. To hypostatise propositions into autonomous essences amounts to setting up a mythology.[39] If a judgment has not been made (as, e.g. the Law of Gravitation before Newton), it is then neither true nor false. Sigwart asks rhetorically:

> Kann man von einem ungeborenen Menschen sagen er gesund ist, weil er gar nicht existiert, ist er dann krank?[40]

Sigwart also asks a question (which I will return to later in this essay), namely, if the process of psychological thinking does not lead to logical cognition how then do we acquire knowledge of logic? Husserl's position leads to two consciousnesses--one, empirical-actual (which must be completely independent of logical laws) and one ideal which apprehends the non-temporal truths. How then could the one reach the other? Husserl, he points out, has a concept of psychology which studies mental life under causal laws and so reaches only vague generalities. It does not analyse self-consciousness and discover in it consciousness of logical necessity. He sets up a "windmill," "gegen die er kampflustig seine Lanze einlegt."[41] And yet Husserl also says that the certainty of logical laws is an "*Erlebnis*." Is not an *Erlebnis* an empirical psychological fact? Is not an evidence a state of the mind which we experience in time?

Natorp in his review of Husserl's book[42] both recognises the value of Husserl's anti-psychologistic arguments and questions whether philosophically one can rest with Husserl's absolute separation between the real and the ideal, the psychological and the logical. First of all, while agreeing with Husserl's conception of pure logic, Natorp questions if pure logic has to be formal. On his view, pure logic has also to be 'material', as Kant's transcendental logic shows. On this issue, one needs to recall that Husserl's pure formal apophantic logic is based on pure logical grammar, and pure logical grammar is not merely a theory of pure syntax, but also has to rule out material countersense such as "Virtue is green," and for this last purpose it has to take into account which material concepts go together and which not. This last must appeal to the way the world is structured, as Husserl comes to recognise in *Formal and Transcendental Logic*. Natorp also takes Husserl to task for finding psychologism in explicitly anti-psychologistic authors, and remarks "One who wants to find the psychological, will find it everywhere, also in Husserl." Just recall Husserl's use of "insight" and "evidence." Natorp of course will reject Husserl's psychologistic reading of Kant, as though Kant's so-called "transcendental psychology (of faculties of the soul)" is also a psychology. But Natorp's main criticism of Husserl is: Husserl leaves the problematic in an unsatifactory state.

After having separated logic from psychology, he still needs to relate the two. For otherwise the material, empirical, psychological, i.e., the real, remains "an uncomprehended, irrational surd."[43] Husserl has to explain, "why is it that the deepest investigation into the constitution of objectivity cannot avoid taking subjectivity into consideration."[44]

SOME RECENT APPRAISALS

Amongst more recent commentators, I will mention only three. On the one side there are Dallas Willard and Barry Smith, both of them use a most interesting reading of Husserl to show why and how, in spite of his use of an act-based theory of meaning, Husserl still succeeds in avoiding psychologism. On the other side there is Michael Dummett who finds Husserl's act-based theory of meaning leading not only to psychologism (which he, in agreement with Frege, rejects), but also as giving rise to several serious problems.

Willard first formulates what he calls "the paradox of logical psychologism"[45] which arises out of two conflicting claims: (i) that the logical truths are about and apply to particular conceivings, inferrings, etc. of particular persons; and yet (ii) that they do not derive their evidence from examination of these mental events of particular persons. Husserl's solution consists in maintaining that such entities as propositions (in general, ideal meanings) relate to particular acts (of judging, for example, in which the same proposition is entertained) as the species Redness relates to particular instances of red color. Propositions then are not objects of acts of thinking, but what Willard calls "complex referential characters or qualities of such acts." The solution of the paradox then is: the logical truths are primarily about, not acts themselves, but certain universal characters of those acts. The truths therefore apply to the acts, and are derived from examination of those acts--while still being strictly universal. In a later essay, Willard writes: "For Husserl, the signification is never the object of the act...rather, it is the intentional character or property of the act or expression.[46] The act is not directed toward the sense but exemplifies it.

Barry Smith develops a similar interpretation of Husserl. (Note that both Willard and Smith prefer the theory of meaning Husserl had in the first edition of *Logische Untersuchungen* and which he abandoned very soon afterwards in favor of his famous concept of the noema.)[47] Comparing the competing strategies of Frege and Husserl for overcoming psychologism, Smith raises two questions for both: first, how do the mind-independent senses relate to the psychological acts of the mind; secondly, how are we able to grasp the senses? His solution to these two questions lies in the way Husserl conceives of the relation of senses to mental acts. Mental acts are complex events which can be sliced into dependent parts in many ways. Apart from such parts as sensory contents and *Auffasungsmomente*, there is also an act-part which is none other than the act's object-directedness. A sense is the universal species which is instantiated in such act-parts (of many different acts directed towards the same object in the same way). The mental acts therefore are necessarily subject to the laws obtaining amongst such species, just as geometrical laws govern real spatial figures.

Dummett[48] is not convinced that Husserl's act-based theory of meaning is able to explain how logical truths are grasped and also how logic applies to mental acts. Recalling Frege's well known exasperation regarding the "mystery" surrounding the idea of "grasping" a thought, Dummett concedes that "it is in taking thoughts as objects of mental acts that Frege goes astray."[49] Frege should have said that they are rather ways in which things are given to us. This is certainly an important improvement upon Dummett's earlier view, namely, that Frege avoids psychologism, while Husserl gets caught in it, because of Frege's attention to language (and Dummett's Wittgensteinian view of linguistic practice) and Husserl's concern with acts. Dummett should have recognised that making meanings dependent upon linguistic practice would have led to what Husserl calls "anthropological" or "specific" relativism. But Dummett now recognises that "where both Husserl and Frege failed was in drawing an absolute line of separation between the psychological and the logical." They thereby "deprived themselves of the means to explain what it is to grasp a thought."[50]

As regards the claim by Willard and Smith that the theory of meaning developed by Husserl in the first edition of *Logische Untersuchungen* solves this problem about "grasping," we should ask: if the act instantiates a sense, how can it grasp the universal instantiated? Husserl's obvious answer is: by reflection. The pre-reflective act is object-oriented, it grasps the sense only by reflecting upon its object-directedness. It does not initially grasp the sense. Here Husserl's formulation was more cautious than Frege's.

PROBLEM OF LOGICAL COGNITION

The problem of what is called 'logical cognition' is not merely the problem of how it is possible to grasp such ideal (or, abstract) entities as propositions (with which Frege was concerned and to which Smith and Willard find a cogent reply in Husserl), but also the problem of how logical truths are apprehended. Husserl was more directly concerned with the latter problem than with the former. It is of course obvious that the latter presupposes the former. The mind must be able to grasp propositions in reflection, before it can apprehend logical truths, for logical truths are either propositional truths or involve propositions. Now with regard to the second problem of logical cognition, perhaps the best account and defense of Husserl's theory is to be found in Robert Hanna's paper "Logical Cognition: Husserl's *Prolegomena* and the truth in Psychologism" (referred to earlier in this essay).

Hanna takes Husserl to have rejected a strong logical psychologism, but not weak logical psychologism. According to weak logical psychologism (which on Hanna's formulation it is correct to say Husserl held),

> (2) (x) (x is a logical proposition------M (∃y) (y is a finite rational
>
> thinker & x is an object of y's thinking [allowing for the changes I introduced in (2) earlier]).

Note that a 'logical proposition' here is not just any proposition or any thought the mind can grasp, but a logical truth. (But one may generalize the thesis to any proposition or thought, and hold that if x is a thought, then it is necessarily possible that there is a finite thinker who thinks it). According to Hanna, Husserl's view is that only non-empirical or rational insight yields the sort of evidence that is rationally conclusive for belief in a logical truth. Experience of self-evidence, as Husserl states in *Prolegomena*, does not guarantee the truth of a proposition.[51] But if a proposition is true, then it necessarily must be possible for some rational thinker to have the experience of self-evidence about it. If the proposition is a contingent factual truth, the experience of self-evidence will be connected with empirical perception, if the proposition is a logical truth then the experience of its self-evidence must be connected with a rational, non-empirical insight. Which is to say, all logical truths are thinkable through rational insight--not that there is actually someone who has rational insight into it. Husserl clearly will reject the empiricist claim that all knowledge is justified by appeal to inductive generalisation on the basis of empirical evidence. This latter thesis is inadequate to account for modal knowledge.

What is important is that the appeal to insight is not falling back on psychologism, i.e., it does not amount to grounding logical truth on a psychological experience. (Nor is grounding empirical truths on empirical perception psychologism, for the same reasons). First of all, it is not to be taken as a theory of truth, but rather as an epistemological theory of experience of truth. So there need be no worry that truth itself is being psychologised. But secondly, the thesis (2) is intended to be a conceptual truth, and not a contingent truth about human minds.[52] It is not intended to be an inductive generalisation.

HUSSERL'S PROBLEMS

Husserl needs to:
(1) have a satisfying way of connecting the logical and the psychological, after the ideality of the logical has been secured;
(2) have an account of our cognition of the logical;
(3) have an account of how the logical truths apply to mental acts of judging, inferring, etc.;
(4) develop a non-psychologistic theory of evidence (and so of cognition and truth); and
(5) be able to overcome naturalism in a more radical way than he does in *Prolegonema*.
(6) He should also take Kant's transcendental psychologism more seriously and be able to show why and how is it that objectivity is constituted in subjectivity, just as there is no subjectivity without objective directedness.

For this purpose, he needs (7) a satisfactory theory about the relation between the transcendental and the empirical. He needs also to argue (8) that the subjective is not *eo ipso* psychological (or even, neuro-physiological), that the psychological is only a certain interpretation of the subjective life of consciousness. Psychologism can be overcome only when the origin of this interpretation is identified.

It is no wonder that in his writings after *Prolegomena*, Husserl searches for a deeper way of both understanding and overcoming psychologism. Such an overcoming would not be merely refuting it by showing its logically (or, pragmatically) incoherent consequences, but by showing the historical origin of such a naturalistic interpretation of the mental life. In the Fifth Logical Investigation,[53] he speaks of "psychological apperception" which has to be overcome through a phenomenological intuition of the essence of the sort of experience that is being investigated. In the Forward to the second edition of the Sixth Investigation, he writes that the "grotesque reproach" that he had fallen back into psychologism shows that the critics had not carefully read the chapter on "The a priori laws of Authentic and Inauthentic Thinking."[54] At this place, Husserl maintains that the laws of logic are the analytic laws of authentic thinking (i.e., thinking whose meaning intention is fulfilled by corresponding categorial intuition of the states of affairs intended) and serve as norms for inauthentic thinking. These laws have nothing to do with 'our' mental organisation. They are grounded in the essences of the act-species and act-forms. "What is seen to be incompatible *in specie*, cannot be brought together and be rendered compatible in empirical instances."[55]

Again in *Formal and Transcendental Logic* Husserl refers to the "bogy of Psychologism." The Logical Investigations, he tells us, had established the objectivity of ideal structures as against psychologically "re-interpreting" them.[56] He makes it clear that in the *Prolegomena* psychologism pure and simple was never thematised. "Rather the discussion concerned a psychologism with a quite particular sense, namely the psychologising of the irreal significations that are the theme of logic. He concedes that at that time he had not entirely overcome "a universal epistemological psychologism." Now he is in a position to radically overcome "the problem of transcendental psychologism."

NOTES

[1] Frank Kirkland, in: D.P. Chattopadhyaya & Frank Kirkland, *Phenomenology East and West. Essays in Honor of J.N. Mohanty*. The Hague: Kluwer, 1993.

[2] G. Frege, "Rezension von E. Husserl", *Philosophie der Arithmetik*, *Zeitschrift für Philosophie und philosophische Kritik*, 103, 1894, pp. 313-332.

[3] M. Dummett, *Frege and other Philosophers*, Oxford: Clarendon Press, 1991, p. 80.

[4] G. Frege, *Nachgelassene Schriften, ed. H. Hermes, F. Kambartel, F. Karlbad. Hamburg: 1969, p. 115.*

[5] G. Frege, *Kleine Schriften*, ed. I. Angeleli. Hildesheim: G.Olms, 1967, p.359.

[6] See esp. Th. De Boer, *The Development of Husserl's Thought*, tr. Th. Plantiga, The Hague: Nijhoff, 1978, esp. pp. 116-7; Th. Seebohm, "Psychologism Revisited," in: Seebohm, Føllesdal, & Mohanty, eds. *Phenomenology and the Formal Sciences*, Dordrecht: Kluwer, 1991, pp. 149-182; Adrian Cussins, "Varieties of Psychologism," *Synthese*, 70, 1987.

[7] Cp. E. Husserl, *Formal and Transcendental Logic*, tr. D. Cairns, The Hague: Nijhoff, 1969, pp. 254-5.

[8] Cp. E. Husserl, *Logische Untersuchungen*, Bd. 1 (*Prolegomena zur reinen Logik*), 2nd. edition, Halle: Max Niemeyer, 1928, secs. 28 and 38. (This book will be henceforth referred to as *Prolegomena*.)

[9] E. Levinas, *The Theory of Intuition in Husserl's Phenomenology*. tr. Andre Orianne, Evanston: Northwestern University Press, 1973, pp. xxxvi, 14-5, 97.

[10] Cp. J. Margolis, *Historied Thought, Constructed World*. Berkeley: University of California Press, 1995, pp.88-9.

[11] R. Hanna, "Logical Cognition: Husserl's Prolegomena and the Truth in Psychologism", *Philosophy and Phenomenological Research* 53, 1993, pp. 251-76.

[12] For this, see D. Føllesdal, "Husserl and Frege: A Contribution to Elucidating the Origins of Phenomenological Philosophy" (tr. Claire Hill, in: Leila Haaparata (ed.), *Mind, Meaning and Mathematics. Essays on the Philosophical Views of Husserl and Frege* (Synthese Library). Dordrecht: Kluwer, 1994.

[13] E. Husserl, *Prolegomena*. pp.129-132.

[14] P. Kitcher, "Frege's Epistemology," *Philosophical Review*, lxxxviii, 1979, pp.235-262.

[15] E. Levinas, loc. cit., p. 15.

[16] Published in *Philosophische Monatshefte*, 23, 1887.

[17] *Logische Untersuchungen*, Bd. 2, Inv. 1, sec. 29.

[18] Ibid, sec. 29.

[19] *Prolegomena*, sec. 24.

[20] Ibid, sec. 24

[21] Ibid, sec. 20.

[22] Cp. G. Frege, *The Basic Laws of Arithmetic*. Berkeley: University of California Press, 1964, p. xvi.

[23] *Prolegomena*, Ch. IV, sec. 24.

[24] *Prolegomena*, sec.35

[25] P. Adler, "Prolegomena to Phenomenology: Intuition or Argument?" in *Graduate Faculty Philosophy Journal*, Vol.16, 1993, no.1, pp.3-76.

[26] Martin Kusch gives a wideranging survey of responses to Husserl on this issue in his "The Criticism of Husserl's Arguments against Psychologism in German Philosophy 1901-1920" in *Mind, Meaning and Mathematics*, pp.51-83.

[27] I. Kant, *Kritik der reinen Vernunft*, B 78.

[28] I. Kant, *Logik*. Ein Handbuch zu Vorlesungen, herausg. von Jäsche (*Kantwerke*, herausg. E. Cassirer, Bd. VIII), p. 334.

[29] J. Herbert, *Sämtliche Werke*, Bd. 4 Langensalza: Hermann Beyer & Söhne, 1891, p.68

[30] E. Husserl, "Entwurf einer Vorrede zu den *Log. Unt.*, 1913," *Tijdschrift voor Filosofie*, 1, 1939, pp.128ff.

[31] For Brentano's alleged psychologism, cp. R. Chisholm, "Brentano's Descriptive Psychology" in Linda McAlister (ed.), *The Philosophy of Brentano*. Atlantic Heights, N.J.: Humanities Press, 1977, pp.91-100, esp., p.98.

[32] W. Schuppe, *Grundriss der Erkenntnistheorie und Logik*, 1st. edition, Berlin: Weidenmannsche Buchhandlung, 1878; 2nd edn. 1910, pp.1-3.

[33] B. Erdmann, *Logik*. Bd.1, 2nd edition, 1907, Halle: M. Niemeyer, p.532-3 fn.

[34] W. Wundt, "Psychologismus und Logizismus" in *Kleine Schriften*, Bd.1, Leipzig: Verlag von W. Engelmann, 1910.

[35] Ibid, p.569.

[36] Ibid, p.579.

[37] Ibid, p.608

[38] C. Sigwart, *Logik*, Bd.1, 4th edition, Tübingen, 1911.

[39] Ibid, p.24 fn.

[40] Ibid, pp.24-5.

[41] Ibid, p.25.

[42] P. Natorp, "On the Question of Logical Method in relation to Edmund Huserl's Prolegomena to Pure Logic" tr. J.N. Mohanty, in: J.N. Mohanty (ed), *Readings on Edmund Husserl's Logical Investigations*, The Hague: Nijhoff, 1977, pp.55-66.

[43] Ibid, p.65.

[44] Ibid, p.66.

[45] D. Willard, "The Paradox of Logical Psychologism: Husserl's Way out," in: *Readings on Husserl's Logical Investigations*.

[46] D. Willard, "The Integrity of the Mental Act: Husserlian Reflection on a Fregean Problem," in Haaparanta (ed), loc. cit., pp.235-262, esp. p.256.

[47] B. Smith, "Husserl's Theory of Meaning and Reference," in Haaparanta (ed), loc. cit., pp.169-183.

[48] M. Dummett, *Frege and Other Philosophers*.

[49] Ibid, p.253.

[50] Ibid, p.256.

[51] *Prolegomena*, sec. 51.

[52] Ibid, sec.50.

[53] *Logische Untersuchungen*, Inv.5, sec.27.

[54] Ibid, Vorwort, Bd.II, II Teil.

[55] *Logische Untersuchungen*, Inv.VI, sec.64.

[56] *Formale und transzendentale Logik*, sec.56.

2. HUSSERL'S THOUGHTS ON THE FOUNDATION OF LOGIC*

INTRODUCTORY REMARKS

It has been one of the strange and unexamined features of contemporary philosophy that philosophical as well as technical concern with the formal sciences — formal logic and mathematics, formal syntax and semantics — has been taken to be an ally of a positivistic, formalistic and (narrowly) analytic philosophical perspective. I believe that upon further questioning, this alleged alliance breaks down, and the prima facie neutrality of the conecrn with formalism leaves room for a great diversity of philosophical standpoints — from a Platonic to a Kantian, to name two familiar contrasting positions. It is also a philosophically unexamined point of view, which requires that an understanding of the nature of the formal sciences must itself be formalistic, that you truely understand the nature of a formal discipline when you embed it in a meta-discipline of the same nature. There is certainly a sense of "understanding" in which this is the case. But there is a sense of "understanding," a sense that is of central importance to philosophy, in which one can ask questions about the formal sciences which do not permit any non-trivially formalised answer. The purpose of the present essay is not merely to bring to the forefront the relevance of the phenomenological perspective for philosophy of the formal sciences, but to open the doors for questioning that taken-for-granted alliance and for the possibility of radically other approaches. Let us only remember that Frege was a sort of Platonist-cum-neo-Kantian (despite the efforts of the believers in that alliance to show that he was a hidden nominalist), David Hilbert a Kantian, Hermann Weyl a sort of Husserlian. Where to place Gödel and Hao Wang? Panini and Chomsky? Or, let us think of our friends Jaakko Hintikka and Carlos Rota.

---

* Originally appeared under the title "Husserl's Formalism" in Følledsdal, Mohanty, Seebohm, eds. *Phenomenology and the Formal Sciences*: The Hague, Kluwer, 1991.

TWO ASPECTS OF HUSSERL'S CONCERN WITH FORM

Looking back at the long and uninterrupted concern of Husserl with formal logic and mathematics, we can say that it has two aspects: a subjective and an objective aspect. The subjective aspect consists in the concern with the mental acts and the operations that lead to the constitution of the forms. The objective aspect concerns the way the forms are related to the world. The first yields, in the long run, a transcendental logic; the second leads to the conception of a formal ontology.

If phenomenology, as transcendental philosophy, is to unravel the constitution of the world, one direct path is to delineate the constraints to which any world whatsoever will be subjected. These constraints are laid down by formal logic as a formal ontology. Formal ontology, in other words, contains the truths which are true in any possible world. Once the formal ontology is laid down, *part* of the task of showing the constitution of the world, of any world whatsoever, will be taken up along with the task of showing how, by what sort of subjective operations, the fundamental concepts of formal ontology come into being. It is therefore not without reason that Husserl returned time and again — every time after a work on transcendental phenomenology — to the theme of formal logic.

If the problem of the world connects with the problem of formal logic — not unlike as with Kant — so does the theory of consciousness, of intentionality, require consideration of specifically those mental operations which constitute formal-logical and mathematical thinking, but also of those which lead to the production of logical forms and mathematical entities. One cannot have an adequate picture of consciousness unless one takes into account these important operations and abilities. There are philosophers — amongst them many phenomenologists — who regard logical thinking and mathematical operations as concealing, rather than exemplifying the intentionality of consciousness. These phenomenologists fail to appreciate Husserl's continuing concern with logic and mathematics. Were that concern only an effect — which time could not wear out — of the early training under Weierstrass, it would not have led, as late as 1927, to the production of *Formal and Transcendental Logic*. The reasons lie deeper. A generalized theory of intentionality needs exploration of all levels of intentional constitution — perceptual as much as logico-mathematical.

THE OBJECTIVE ASPECT

In the phenomenological inquiry, there is a methodological primacy of the objective noematic description. It is only after the noematic, constituted sense has been laid bare, that philosophical thinking can return to those constituting operations which make such sense possible. So let us first consider what formal logic and mathematics — Husserl was convinced of their inner unity — are about, what the very sense of their being consists in, and what their *telos*, the goal toward which they grow, is.

1.The purpose of logic is the construction of a *theory of deductive sciences*. Logic is not limited to a theory of purely formal consequences or to a theory of logical inference, although any such theory would find a place within logic. As a theory of deductive sciences, its task is to exhibit the form of all deductive theories, the form of which all specific theories are specifications. In logic, therefore, all material concepts have to be left undetermined, and replaced by some specification of the formal concept of "object-in-general." Even the modes of connection have to be determined exclusively by their form, i.e., by the form of elementary laws taken to be valid of them.[1] Thus in logic "+" is not a sign for addition of numbers, but for a connection for which, amongst others, laws of the form "a+b=b+a" hold good.

2. Husserl arrived at this position by following two paths. In *Philosophy of Arithmetic* (1891), he claimed to have arrived at the conclusion that the *formal* in the mathematical disciplines lies in relatedness to *"objectivity in general," "something in general,"* in a most empty universality which leaves every contentual determination undetermined. Formal mathematics, in this sense, is to be clearly distinguished from its application in any *real* mathematics such as number theory, theory of functions or Euclidean geometry. In an essay "Arithmetic as Deductive Science" of 1885, Husserl writes that the concept of formal number is the mere concept of *something* whose objects underlie connections and relations and laws which are expressed in the appropriate axioms.[2]

Also around the years 1887-1895, Husserl was engaged in developing a philosophy of purely formal, calculatory operations — by which he wanted to solve two specific questions that worried him: one concerned "imaginary number" and the other the extension of number systems. What he did was to understand all arithmetical calculation as mechanical, purely symbolical and formal operation, as "a systematic deduction of symbols from symbols in accordance with rules for those symbols."

These two paths matured, under the influence of Hilbert, into an axiomatic conception of theory, just around 1901.

3. According to this mature conception, by determining the theory-form, or the form of any deductive theory whatsoever, logic and formal mathematics determine a domain of purely formal objects — the so-called *Denk-objekten* of mathematics — purely by virtue of the axioms of the theory. This gives rise to Husserl's idea of a definite manifold.

A manifold is a domain of objects. The nature of this domain is completely determined by the sole fact that it remains under a theory of such a form. Objects belonging to this domain, though totally undetermined as to their matter, are totally determined by the forms of their interconnections which themselves are determined by the forms of elementary laws taken to be valid of them. Euclidean geometry, by its *form*, defines such a domain of objects, no longer to be understood as spatial figures. Euclidean geometry itself is a *real* mathematics, but when formalized, it is shown to be specification of a theory-form — the latter determining a domain of Euclidean manifold. A further formalisation to a higher level yields what would be the form of of which these theories are specifications — namely, the form of all possible manifolds. Such a formal manifold theory would, however, still be using formal-ontological concepts such as one/many, object/property/relation, order/sequence, identity, and so is still an ontology — indeed a purely formal ontology.

4. Whereas pure mathematics grows toward such a formal ontology (and so has to talk about formal objects — such as classes/sets, sequences/orders, relations and structures), formal logic as a logic of predicative propositions i.e. as apophantic logic still remains basic in the sense that any theory, even the pure theory of theory-forms, has to be a valid concatenation of propositions — more accurately, of propositional forms.

5. The dual nature of formal logic — on the one hand, it is a logic and on the other it is formal ontology — was of great importance for Husserl. The general argument on which this thesis is based is simple:[3] if logic is concerned with *meanings* (i.e. concepts, propositions, theories), every meaning refers to an *object*. To specific meanings (in specific sciences), there correspond specific objects. To formal meanings (such as "object," "predicate," "relation"), there correspond formal objects: concept, class, sequence, law, etc.

6. Formal logic is as much different from real logic (i.e. logic of specific sorts of knowledge, or of specific regions of *reality*), as formal ontology from *real or ideal* ontologies (*a priori* or *a posteriori*), i.e. from regional, material ontologies.[4]

7. Formal logic *qua* logic (i.e. not regarded as formal ontology) itself exhibits an internally stratified structure. At the ground level, pure logical grammar lays down the rules for avoiding nonsense, i.e. rules that determine which concatenations of expressions are well-formed sentences. These are syntactic rules of formation. At the second level, the logic of analytic consequence lays down rules for determining which compound well-formed sentences are consistent, which self-contradictory, and which necessary. What is important for Husserl, at this second level, is that we still do not introduce the semantic predicates "true" and "false." The laws of analytic consequence are to be formulated independently of these semantic concepts. At the highest level, one introduces the concepts "true" and "false," and lays down the laws that determine their correct applications.

Although in the earlier discussions of the subject matter, Husserl did not clearly state whether there are formal ontological correlates of the first two strata of pure logic, in *Formal and Transcendental Logic*, he clearly takes a stand.[5]

He asks: Why is not the pure analytic of mere non-contradiction (i.e. the second stratum of pure logic) already formal ontology, if it is concerned with judgments and judgments are concerned with something in general? To this, he answers: The pure logic of non-contradiction is at most an ontology of forms, of consistent and so possible meanings. The possibility of a meaning does not *eo ipso* guarantee the possibility of objects corresponding to that meaning. The question of possible *truth*, of relation between consistent meanings and possible objects goes beyond the pure theory of meaning. As long as we are concerned with judgments (or judgmental forms) as our exclusive theme, we are not yet doing formal ontology proper. For logic to become ontological, the interest must switch from concern with meanings to concern with objects and so with truth.

I will press the point that this position raises problems that cast doubt on the entire concept of formal ontology. But before turning to the critical remarks let us, as earlier indicated, turn to the subjective approach.

THE SUBJECTIVE ASPECT

(a) The Psychological

In his 1906-1907 lectures on logic and theory of knowledge,[6] after developing his concept of pure logic, Husserl asks: even if we leave out of consideration the historical facticity of the researcher's psychological states, can we completely leave out such references to intellectual acts as are implied in locutions like, "Given this, is it justified to ask, or to suppose that...?" Or: "If one is sure that the proposition A holds good, one cannot any more doubt whether the proposition B holds good or not."

These locutions do not express merely mathematical or logical states of affairs, but rather something about our judging, about our certainty, about our questioning, supposing, etc. What is important is that, they all, in this context, concern the problem of justification. One says, e.g. "We do not know whether A holds good, but on the basis of the already developed theory we have 'reasons,' or are 'justified,' to suppose that it most probably does hold." This sort of locution occurs not merely in the empirical sciences, but also in the pure, a priori disciplines.

What we find, then, is that a subjective structure of justifying acts accompanies the logical concatenation of objective propositions constituting a theory. The objectively oriented theoretician performs those acts of justification, but is not thematically directed towards them. His focus is on the objective structures alone.

One needs then a new discipline whose task it is to consider the epistemic acts with regard to their relative justifying claims. Husserl calls it the "Normlehre der Erkenntnis" or better still, simply "Noetik."[7]

Noetically, formal logic, like any other objectively focused science, proceeds with a naivity. One needs therefore a reflective switch of interest, which shall amply complement the objectively directed logic. Such a *noetic* would consist in an eidetic psychology, and some of the structures it would countenance are like this:

For any knower S, if S has performed an epistemic act of type Ψ at time t_1, then he can*not* perform an epistemic act of type Φ at time t_2, if both these acts are to have the same propositional content.[8]

(b) Meanings as Intentional Contents:

That the subjective acts are not mere contingent accompaniments of the objectively oriented logic comes to light, when we consider that objective logic is concerned with such entities as propositions or propositional forms which are meanings, and meanings are but contents of intentional acts such as asserting, denying, believing, assuming, etc. They are contents which can be shared by numerically as well as qualitatively distinct acts. The objective logician abstracts the identical contents from the act-context which is their original home. The phenomenologist respects that abstractive focusing, but wants to reflectively overcome the naivity that is unaware of the act of abstraction.

(c) The Constituting Acts:

There is, in fact, a double abstraction involved. There is first the abstraction of the content from the act — the content being a "material" or *sachhaltiges* proposition, syllogism or theory. Then there is the formalizing abstraction which replaces the material terms by empty variables. The two acts of abstraction are not merely necessary accompaniments of formal-logical thinking, they are essentially constitutive of the entities formal logic deals with. Note that the "material" proposition on which formalizing abstraction is exercised need not be an ordinary, empirical proposition such as "snow is white," but may be a proposition from pure arithmetic, number theory or non-euclidean geometry.

(d) Formalization and Generalization:

It is important for Husserl's understanding of "form" that the process of formalization be sharply distinguished from the process of generalization. The two are *radically* different sorts of intellectual acts. By generalization, we move from species to genera and thus to the highest genera such as "material things," "living beings" or "mental beings." If the species is defined by giving the genus and the differentia, the genus is a part of the species, taken in *intension*, while the species is a part of the genus, taken in *extension*. But the concept "object in general" is *not a genus* under which such species as "number" and "material objects" are to be subsumed. This concept is reached not by greater and greater generalization, but rather by more and more formalization. The schema "$p \supset q \cdot p : \supset q$" is not the genus of all instances of it, but is their form.

Hence the necessity of drawing a sharp distinction between material essences and formal essences. Material essences can be arranged under the highest regional essences constituting regional ontologies. But the formal essences are but categories of the empty region — and so a quasi-region of "object-in-general." The category "essence" itself is a formal essence. The essence "red" is a species of the essence "color," but not a species of the essence "essence," it is rather an instance of it. The highest formal essence "object-in-general" differentiates not into other species, but into *categories* such as "property," "relation," "group," "number," "whole-part," "genus-species," "sequence." These are formal-ontological categories.

By keeping the two — formalization and generalization — apart, and, consequently, by sharply distinguishing between the formal ontology and the regional ontologies, Husserl may be said to have avoided several ontological alternatives: most importantly any *reduction of* "material" entities (and their essences) — including numbers — to purely formal structures. If he was a platonist with regard to numbers and such abstract entities, he was also a formalist, a follower of Hilbert, at the level of pure *formal* mathematics and so of formal theory of manifolds and formal ontology.

Problems:

Husserl's theories, which I have briefly sketched above only in some of their salient features, give rise to several problems.

To begin with the last-mentioned point: to what extent Husserl was a formalist and to what extent he was an intuitionist? How is his platonism with regard to such ideal entities as numbers compatible with his formalism? At this point, interpretation is very difficult. Some like Hermann Weyl regarded him as an intuitionist and constructivist out and out. Others have noted his undoubted closeness to Hilbert. As far as I can see, with regard to "material" or contentual mathematics — such as number theory and function theory — he was a constructivist and intuitionist, requiring that empty existential statements need to be filled in by intuitive i.e. constructivistic evidence; with regard to *formal* mathematics he was a Hilbertian, committed to an axiomatic structure, and to the ideas of consistency and decidability. In fact his position appears to be *similar* to that adopted by Hilbert in response to Brouwer. Hilbert adopted intuitionism in one part of mathematics and formalism in another. Husserl, I think, retained intuitionism in what he regarded as real i.e. *sachhaltiger* mathematics, but regarded formal mathematics to be axiomatic and formalistic in the sense that all reference to the meanings of the signs is to be dispensed with. When Hermann Weyl writes in his obituary essay on Hilbert that

> An impartial attitude will do justice to both sides; not a little of the attractiveness of modern mathematical

> research is due to a happy blending of axiomatic and genetic procedures.[9]

he was, I think, confirming the basic insight of Husserl.

How is all this to be reconciled with his Platonism? I think Rick Tieszen has nicely shown that his intuitionism and constructivism make room for the transcendence of what is so intuited and constructed: what one needs to do is to stress the idea that the series of mental acts which *bring an object to intuitive givenness* is a law-like sequence fixed in advance by a rule, and, in the case of large numbers not actually completable — which would be a way of capturing their *transcendence*.[10]

Did Husserl accept the law of excluded middle? It would appear from remarks taken out of context as though he wanted to cast doubt on the validity of that principle.[11] However, that may be misleading. While laying bare the inner nature of formal logic, he regards formal logic as committed to decidability and the principle of excluded middle. (That principle, like the other logical principles, has for him two forms: one belonging to the logic of pure consistency, and the other to the logic of truth.) The task of transcendental logic is to critically reflect on this, as on the other naivities of formal logic — not with a view to reject some of them if necessary but to understand their *origin*, and range of validity.

Husserl's idea that pure formal mathematics culminates in a theory of *definite* manifolds expresses a pre-Gödelian optimism that was shattered in 1930. For such a domain, he writes, as late as 1927, there is no truth that is not analytically contained in the axiom system and every sentence that can be logically-grammatically constructed out of the fundamental concepts is either "true" (i.e. an analytic consequence of the axioms) or "false" (i.e. an analytical contradiction).[12] I do not know of any material in the *Nachlass* in which he shows familiarity with and response to the Gödelsatz.

In these regards many of his theses will be subjected to the same critical judgments as Hilbert's are. But in many ways, philosophically, his position is more comprehensive and multi-faceted, and so has better chances of withstanding researches in the field.

Kreisel has said that "what characterises the difference between e.g. the idealist and the realist is what aspects of (crude) experience (in this case, mathematical experience) are regarded as significant and suitable for study," and has contended that when idealism and realism are sharply formulated as mathematical theses it may turn out that they do not exclude one another.[13]

I think this is exactly the philosophical point that emerges if we take Husserl's work as a whole, which shows a struggle towards making a Platonism consistent with a constructivist intuitionism.

Feferman in a paper entitled "Categorical Foundations and Foundations of Category Theory" makes a general point that is directly relevant to understanding Husserl's motivation:

> Since neither the realist nor constructivist point of view encompasses the other, there cannot be any present claim to a *universal foundation* for mathematics, unless one takes the line of rejecting all that lies outside the favored scheme. Indeed, *multiple foundations* in this sense may be necessary, in analogy to the use of both wave and particle conceptions in physics.[14]

It seems to me that a truly phenomenological approach towards logic and mathematics cannot but lend support to such a point of view.

These positive evaluations should not, however, blind us to the many difficulties that affect Husserl's ideas — especially his idea of formal ontology. As regards the conception of formal ontology (leaving out the contradiction which Gilbert Null has recently found in the theory[15]), I will here mention three problems. In the first place, I find a purely syntactic — as opposed to a semantic — conception of formal-ontolological categories unsatisfactory. Husserl, in this context, regards the concept of "object" as a purely syntactic category. If by "syntactic" category is meant what relates to the syntactic role of an expression (object is that which is designated by a syntactically nominal expression), then you need not only expressions which designate objects, but also those which by virtue of their syntactic role do not. For example, in Frege's semantics, predicates designate concepts which are not objects. For Husserl, concepts are also objects, so also are relations and functions. His idea of *syntactic* notion of "object," then, is vacuous. My second anxiety is, how this syntactic idea of "object" is to be fitted with a *semantic* idea of object which Husserl has, namely that which follows from his thesis of intentionality. Thirdly, Husserl holds in *Formal and transcendental Logic*, as we have noted, that formal ontology corresponds to logic of truth and not to a logic of consequence. If the latter is concerned with meanings (such as concepts, propositions, theories), and if meanings not only refer to objects but are themselves objects, irrespective of which meaning-structures are *true* (in the actual world), a logic of consequence will also yield a formal ontology of Husserl's sort. But meanings being semantic entities, such a formal ontology will be ill at ease with a purely *syntactic* idea of that discipline. The fact is that Husserl did not clearly distinguish between syntax and semantics, but no one of his contemporaries did.

One must also add that logic as formal ontology, insofar as it is founded upon a pure logical grammar, refers back to a "material" ontology which would have to be the source of those "material" incompatibilities (or, compatibilities) which pure logical grammar "overlooks" insofar as "Virtue is green" or "2 is blue" is syntactically well-formed, but yet is not a possible actual judgment. A semantic theory of which categories of meanings are "compatible" and which "incompatible," if it is not to be arbitrary, must refer back to the way things are in the world as experienced.

None of these however is an irremediable problem. I must add that one has to distinguish between Husserl's own views about logic and mathematics and what a proper phenomenological view should be. I think a phenomenological philosophy has conceptual resources that would be enormously helpful to achieve the kind of reconciliation that the quotes from Kreisel and Feferman suggest. I heve particularly in mind: the concepts of "evidence," "intuition" (tied to the ideas of "construction," "fulfillment" and sequential ordering of perspectives), "constitution" (which is neutral as against realism and idealism), "founding — founded relation" (which recognizes different levels and strata of objectivities) and "transcendental subjectivity" (which appropriates the role of mental acts in constitution without falling into an unwanted psychologism). In this essay I will not attempt to show how these concepts help in resolving some apparent tensions in philosophy of mathematics. What I have intended to do is to indicate both the line of that approach and that some current researches point in that direction. In this highly muddy field that is the best that I could hope for.

NOTES

[1] E. Husserl, *Formal and Transcendental Logic*, E. tr. D. Cairns, The Hague: M. Nijhoff, 1969, §28.

[2] E. Husserl, *Studien zur Arithmetik und Geometrie*, ed. I. Strohmeyer, Husserliana Bd. XXI, The Hague: M. Nijhoff, 1983, p.66.

[3] E. Husserl, *Einleitung in die Logik und Erkenntnistheorie*, ed. U. Melle, Husserliana Bd. XXIV, Dordrecht: M. Nijhoff, 1984, §14, pp. 51 f.

[4] Ibid, ch. 3.

[5] *F. u. t. l.*, §54.

[6] Husserliana Bd. XXIV, ch. 4, esp. §25.

[7] Ibid, p. 134.

[8] For more about this, see my "Psychologism in Indian Logical Theory," in Matilal & Shaw (eds.), *Analytical Philosophy in Comparative Perspective*, Dordrecht: Reidel, 1985, pp. 203-211.

[9] H. Weyl, "David Hilbert and his Mathematical Work," Reprinted in: C. Ried, *Hilbert*, New York: Springer Verlag, 1970, pp. 245-283, esp. p. 274.

[10] R. Tieszen, "Mathematical Intuition and Husserl's Phenomenology," *Nous*, 18, 1984, pp. 395-421.

[11] Cp. *F. u. t. l.*, §§ 77, 79.

[12] *Ibid*, p. 100.

[13] G. Kreisel, "Mathematical Logic" in: *Lectures on Modern Mathematics*, vol. III, ed. T. L. Saaty, 1965, p. 186. Also see J. M. B. Moss, "Kreisel's Work on the Philosophy of Mathematics," in R. O. Gandy and G. M. E. Yates (eds). *Logic Colloquium '69*, London/Amsterdam: North-Holland, 1971, pp. 411-438, esp. p. 422.

[14.] in: R. E. Butts & J. Hintikka (eds), *Logical Foundations of Mathematics and Computability Theory*, Dordrecht: Reidel, 1977, pp. 149-169, esp. p. 151.

[15.] Cp. G. Null, "Husserl's Theory of Essence" in: J. N. Mohanty and William McKenna (eds.), *Husserl: A Textbook*. Washington, D. C.: University Press of America & The Center for Advanced Research in Phenomenology, 1989.

3. ASPECTS OF HUSSERL'S PHILOSOPHY OF LOGIC A PROPOS HIS LOGIC LECTURES OF 1906-7*

I propose to preface this lecture on Husserl's philosophy of logic by a few preliminary and elementary remarks on the problems in philosophy of logic which concerned Husserl and on the deeper philosophical reasons for that concern.

To say that Husserl was concerned with the problems regarding the foundation of logic is not wrong, but it is misleading. There are well recognized foundational problems which did not concern him. More specifically, there is a discipline called 'Foundation of logic' (and of mathematics) which itself is a formal (and mathematical) discipline of a higher order. There is a widely held view according to which this is the only way one could hope to be able to make substantial progress in such matters. But it is easy to argue that this view is mistaken, that, in the long run, one must face up to certain *philosophical* questions concerning the foundations, questions which are not themselves formal-logical.

I think, Husserl was concerned with the following groups of questions:

I. First, questions as to the *nature* of logic.

II. Second, questions as to the *clarification* of the fundamental concepts of logic.

III. Third, questions as to the *role of logic in the cognitive process*, and finally,

IV. The question regarding the *relation of logic to the world*, to *Being*.

Let me comment briefly on each of these in order to emphasize, in each case, the peculiarly Husserlian slant in the very questioning.

I. Regarding the nature of logic, there are *two extreme* points of view that Husserl avoids. There is, at one end, the traditionalist who focuses exclusively upon the intention underlying the historical genesis of the concept of logic (of the concepts of logic, thought, reason, etc.) and, as a consequence, considers e.g. the formalization and mathematization as an emptying of that sense (*Sinnesentleerung*), a regrettable historical process which raises doubt upon the very application of the name 'logic' to those new developments. At the other

* Originally appeared in S. IJsseling (ed.), *Husserl-Ausgabe und Husserl-Forschung*, Dordrecht: Kluwer, 1989.

end, are the modernists who are so impressed by the formalization and mathematization of logic, that they refuse to recognize the decisive significance of the original genesis, of the *Urstiftung* which for them is no more than a historical, possibly, a linguistic accident which cannot hold in ransom all present and future developments. It is to Husserl's credit that he recognized *both* that mathematization is a genuine fulfillment of the sense of 'logic' *and* that that sense was implicit in the historical genesis of 'logic'. Both those who deny mathematization and those who are unaware of the decisive role of the original sense, do not grasp the total sense of what logic is about. It is this *total* view which leads unavoidably to the conception of logic as formal ontology.

II. The fundamental concepts of logic are expressed in the so-called logical terms: in the predicate expression 'is (or is not) Φ', in the terms signifying quantification (such as 'all', 'some', 'none'), in the definite or indefinite descriptive phrases ('The Φ' or 'A Φ'), in the expressions for disjunction ('either-or'), hypothetical judgments ('If-then'), negation ('not-p'), and, in the long run, in the concepts of 'proposition' and 'inference'. To *clarify* these concepts, in an ordinary sense, is to precisely state what they mean or do not mean, to recognize different usages of these (such as between exclusive and inclusive disjunction, between term-negation and sentential negation) where they might be confused, and to introduce useful definitions in order to fix their intended use. There is also a sense of 'clarification' in which to clarify a concept is to 'construct' it in intuition, instead of using it merely 'blindly', emptily or symbolically, to relive the experience in which the intention embodied in a concept is fulfilled precisely in the manner, and within the limits that are permitted by that intention. Thinking in logic (as in mathematics) had to be carried on emptily, mechanically, following an algorithm; but philosophical clarification has to lay bare the intuitive formation of those concepts as well as the process by which that living intention needs to be covered up so as to permit an empty, mechanical manipulation. Husserl's concern, in clarifying the fundamental concepts of logic, was precisely to lay hold of the origin of those concepts in the living experience of thinking. Here, again, it is incumbent upon him to avoid *the two extreme errors of psychologism and Platonism* — of the reduction of logical entities to the psychological process of thinking and of the reification of those into self-subsistent 'eternal' entities.

III. There are not two kinds of thinking: the thinking that is involved in our cognitive enterprise (of everyday life and of the sciences) and the thinking that goes on within formal logic. As Kant put it most pertinently, it is the same *Verstand* by the same act which produces on the one hand the analytical forms of judgment and on the other the synthetic forms of transcendental logic. A philosophy of logic should be able to identify this 'sameness' of thought, as operative in the cognitive enterprise that is motivated by the goal of *knowing*, and as reflectively objectified in the work of the logician who turns away from that original cognitive goal in order to focus upon the nature of the thought structures involved in the cognitive process. Logic is formalized, axiomatized, mathematized and embedded into wider formal structures, it assumes an air of autonomy that is sustained by a 'forgetfulness' of its original sense. Philosophy of logic should restore that awareness, *not in order to reverse that process,* but to make logic conscious of its 'origins' — historical as well as cognitive.

IV. Once the cognitive role of logic — as contrasted with its internal, formalistic nature — is borne in mind, one will not any longer want to set aside the questions regarding the relation between logic and reality. Again, as in the case of the preceding questions, one needs to avoid *two onesided and extreme positions*. One of these is the position which starts in the first place with a conception of reality which by definition removes the latter from the reaches of logical thinking, and secondly, with a conception of logic as a *tool* which one *finds* at his disposal and *employs* for a task, i.e., for knowing that reality for which it is ill suited. Both these are unwarranted preconceptions. At the other extreme is the view that reality is logical in its structure, and thought and reality coincide in their outermost reaches if not at a proximate distance — again, a metaphysical preconception that derives its plausibility from *either* suitably transforming logic (so that it ceases to be a formal discipline) *or* appropriately construing what reality is (so that it is stripped of its sensuousness, and reduced to a conceptual structure), *or* from both these steps taken together. *Logic is not a tool* that thought employs. It rather exhibits the structures of thought that are involved as the latter pursues its cognitive goals. Reality need not be construed as being 'on the other side' of thinking, but rather as the object that is being intended by thought in and through various predications, negations, implicative and disjunctive judgments, inferences, theory constructions and confirmation and disconfirmation of hypotheses: this includes everyday life, science, as well as philosophical thinking. In each case, wherever thinking posits an object in whatever modality of being (i.e., as real, as possible, as imaginary, as presumptive, etc.), the object must conform to certain logical requirements. Logical forms are *also* forms of objects in general. Formal logic yields a formal ontology. Formal logic, however, cannot pre-legislate the content of reality. In Kant's language, the content cannot be *a priori* anticipated. The material regional ontologies lay down structures to which any thing whatsoever must conform depending upon to which region it belongs. But these are hypothetical structures: if something is Φ, it must be Ψ. But that it is Φ is contingent, and no material ontology can *a priori* anticipate it. Thus logic does not exhaust reality. At the same time the real must satisfy the *a priori* conditions of formal ontology, so of formal logic.

The deeper motivations of Husserl's continuing concern with logic become now clearer: from the subjective point of view, it was to understand the nature of human thinking within the overall context of a theory of intentionality; from an objective point of view, it was to understand the *formal* structures of the *world qua world*. Both the perspectives need a little emphasis and clarification so as to avoid misunderstandings.

Consider the subjective point of view first. Again, there are two views, widely held, which I want to set aside. Many philosophers hold that a concern with intentionality should lead in the direction of consciousness's directedness towards the world (or man's being-towards-the-world) and so to the multifarious concrete intentionalities in their intricate interweaving: such a concern has no place for concern with logic. Logic deals with abstractions, intentionality with the web of concrete experience. In my view, this way of separating logic from intentionality is fundamentally mistaken. If the entire mental life is characterised by intentionality, so is the life of thought. You do not grasp intentionality in all its aspects if you do not understand the intentionality exhibited in logical thinking (as also in the higher order construction of formal systems) in its relationship to the other levels of intentionality (such as the perceptual). There is a deeply rooted prejudice to the effect that perception is concrete while thinking is abstract. Every *Erlebnis*, every intentionality — be it perceptual or logical — is concrete. Now just as there is this tendency to exclude the logical from the intentional, so there is the opposite tendency to suspect all intrusion of the intentional as psychologistic, and so to reify the logical into an autonomous domain of selfsubsistent (formal) essences.

A brief remark about the objective point of view: Husserl's view that formal logic yields a formal ontology which gives the form of any world whatsoever is different on the one hand from the Kantian view that the transcendental logic yields the most general concepts as applied to *a priori* intuitions whereby the world as an ordered whole is constituted. Kant, in spite of the close relation he forged between formal logic and transcendental logic, did not have the idea of a formal ontology whose domain is the empty region of object in general. This also accounts for the difference between Kant's and Husserl's conceptions of a transcendental logic — but to this I propose to return later in this essay.

II

With these preliminary remarks, let me turn to Husserl's 1906/7 lectures on Logic and Theory of Knowledge (*Hua Vol. XXIV*). In these lectures, Husserl both reiterates the fundamental position about pure logic as a formal theory of science, indeed as a formal theory of all theory-forms, which he had developed in the *Prolegomena*, and introduces two significantly novel themes. I will not pause to elaborate on the well-known conceptions of 'pure logic' as a formal theory of theory-form, of the unity of formal logic and formal mathematics, and eventually of a theory of manifolds. I will rather comment on the three new themes he develops: These are: (A) the discussion of the relation between *formal logic* and *real logic*, (B) the idea of a *Noetik* — as he calls it — i.e., a theory of justification (*Rechtslehre*) of knowledge, and (C) finally, the idea of *philosophical logic*. The first two correspond to the third and the fourth questions in philosophy of logic, as they were stated in my introductory remarks: the question about the role of logic in the cognitive process, and the question about the relation of logic to the world. The third corresponds to Husserl's later idea of transcendental logic.

A. Granted that any reference to the historical facticity of scientific research (i.e., to who proved what and when or out of what motivation) is inessential from the logical point of view, there is however a subjective discourse which, even if it does not refer to historical or psychological facts about researchers, does make use of interconnections of justification amongst mental or cognitive acts. One asks for an example:

'Given this, is it justified to ask, or to suppose that...?' or 'If one is sure that the proposition A holds good, one cannot any more doubt whether the proposition B holds good or not'.

These locutions do not express merely logical states of affairs, but rather something about our judging — about our questioning, supposing, doubting, etc. They all, in this context, concern the problem of justification. One says, e.g., 'We do not know if A holds good, but on the basis of the already developed theory we have reasons to suppose that it most probably does hold'. This sort of locutions occur not merely in the empirical sciences, but also in the pure, *a priori* disciplines.

What we find is that a subjective structure of justifying acts accompanies the logical concatenation of propositions constituting a theory. The objectively oriented theoretician performs those acts of justification, but is not thematically directed towards them. One needs then a new discipline whose task it is to consider the epistemic acts with regard to their relative justifying claims. Husserl calls it the '*Normlehre der Erkenntnis*' or, better still, simply '*Noetik*'. Such a *Noetik* would consist in an eidetic psychology, and some of the structures it would countenance are like this:

> For any knower S, if S has performed an epistemic act of type A at time t_1, then he *cannot* perform an epistemic act of type B at time t_2, if both these acts have the same propositional content p.

When we construct a theory, we live in subjective acts, we live in the performance of conviction and supposition, we experience with insight their justification. But we are not occupied with these subjective experiences, we do not enquire into them. While living in these subjective acts, we rather focus on their meaning contents and their objective contents, we follow the matters themselves and their interconnections. Thus theory requires the justifying acts, but does not enquire into them. Formal logic itself is not the theory of justification (*keine Lehre von den Rechtsquellen*). *Noetik* is the theory of the intellectual attitudes with respect to their justifying claims.

What is the relation between formal logic and *Noetik* (or epistemic logic)? Husserl formulates their relation thus — the two stand in the closest connection. Formal logic can be worked out independent of the epistemic logic, but not *vice versa*, inasmuch as epistemic logic which is a theory of justifying evidence must include the formal conditions of the possibility of evidence. Although formal logic can be 'noetically naive', if it is to be critical and so philosophically sophisticated it should be in close relation to epistemic logic.

B. Husserl also speaks of a *real* logic. If formal logic is theory of theory in general, real logic is (also *a priori*) theory of all knowledge about reality. Real logic then would be concerned with the *a priori* concepts valid of any reality whatsoever, with truths that relate to the categories of reality (such as: 'thing', 'property', 'real connection amongst things', 'real whole' and 'real part', 'cause', and 'effect', etc.).

At one point in his exposition of 'real logic' (*Hua, XXIV* p. 112/113), Husserl recalls Kant's distinction between formal logic and transcendental logic.

Kant, as it is well-known, writes that whereas (general) formal logic abstracts from all contents of knowledge, and considers only the empty form, transcendental logic abstracts only from all empirical contents but is concerned with the relation of thought *a priori* to pure contents, i.e., with the *a priori* thought of *real* objects, or with the Kantian categories in their relation to pure intuitions of space and time. The Kantian transcendental logic, it should be obvious, comes close to Husserl's real logic, but — Husserl cautioned us — the two should not be identified. To pursue this distinction between the two would be to narrate a separate story: one needs to go into their differing conceptions of formal logic (and of mathematics) and their differing conceptions of space and time.

A discussion of the relation between formal logic and real logic becomes the occasion — one of the few occasions in Husserl's writings — to develop his thoughts on the *relation between logic and metaphysics*, and in that context, to tell us *what he means by 'metaphysics'*. His thought traverses the following path.

If to formal logic there corresponds a formal ontology (according to a celebrated Husserlian thesis), to real logic there corresponds a *metaphysical ontology* (p. 95). As Husserl uses this name, metaphysical ontology is concerned with *reality*. Although none of the empirical sciences about the world gives a final knowledge of its domain, they all presuppose, make use of, but cannot, within their limits, clarify the fundamental concepts and principles of 'Reality'. These are for example 'cause' and 'effect', 'thing' and 'quality', 'space' and 'time'. It must be possible to have a 'final' science of reality — Husserl calls it *Seinswissenschaft*, but it means science of real being — which is science of Being 'in the absolute sense'. Husserl also calls it 'metaphysics'. Metaphysics, then, relates to the other sciences of reality, and *presupposes* them (I believe, Husserl means, in order of knowledge, not in order of being). This metaphysics is 'radical ontology' (p. 99). In this widest sense, metaphysics will enquire into what is real in the final sense, it will provide the final interpretive framework for the empirical sciences. Such an *a priori* metaphysical ontology will include e.g., an *a priori* theory of time, and a pure geometry of space. With these we still do not acquire any knowledge of that reality which actually exists, for the *a priori* metaphysics will only tell us what conditions any possible reality must have to satisfy.

From this *a priori* metaphysics, Husserl distinguishes an 'empirically founded metaphysics' which would determine what the most general features of actual reality are: this sort of determination has to depend upon the empirical sciences.

When Husserl asks, what is the relation between logic and metaphysics, he is asking what is the relation between formal logic (and its correlate, formal ontology) and *a priori* metaphysics. He seems to me to be defending the following three theses:

(1) There is a sense in which formal logic (and, correspondingly, formal ontology) is the foundation of metaphysics, for what belongs to being as such, to the empty region of object in general, must also belong to all reality.

(2) But there is also a sense in which formal logic presupposes metaphysics. This dependence can be brought out in various ways. If we consider apophantic logic, every proposition contains *terms* and predicate words. These terms are, to begin with, categorially formed contents (such as 'The man in the grey suit' and 'The apple tree over there'), which, on their part, refer back to the 'stuff' in the absolute sense (which are *not* categorially formed). In this way everything logical points to something extra-logical, to the world of *hyle* — which is the domain of metaphysics. Or consider formal ontology: its widest category is 'object in general'. This category and its logical modalities refer back to the category 'individual' and its logical modalities. Every theory, even *a priori* theories including pure arithmetic, even pure theory of manifolds, has its ultimate singularities. Thus the number 2 is one such in arithmetic. But 2 is also a universal which is instantiated in 2 apples, 2 men, 2 propositions, or even 2 numbers. We thus see that the ideal singularity 2 may relate either to ideal entities or realities as units. But, in the long run, there is always the possibility — *a priori* — of relating to real individuals. All ideal singularities contain this possibility *a priori*. In this way, all thought about idealities refers back — as an *a priori* possibility — to thought about *possible* real individuals (*but is not reducible* to the latter). Or, again, consider the very idea of form. Form *points to* stuff — in the first instance or proximately, to relative stuff, but in the long run to absolute stuff or matter. *Even if no specific reality* is implicated in any form, there is *a priori* a reference to some possible reality. Absolute stuff belongs to the domain of sensuous reality. In logical thinking, pure stuff is left out and replaced by an indeterminate variable — thereby making possible a consideration of forms themselves, and so formal logic. This leads to Husserl's third thesis regarding the relation of formal logic and metaphysics.

(3) In spite of this *a priori* reference *back* (which is *not* a dependence in the usual sense, but an *intentional* dependence) to reality, forms constitute a closed domain of their own into which nothing metaphysical enters.

C. I will now briefly turn to the idea of a philosophical logic which Husserl introduces in these lectures. But before I do that, it is important to point out that there is, in Husserl's mind, a deep ground of the close relation between Noetik, formal logic and real logic. That relation is but a consequence of the way the three aspects of a cognition hang together: these three aspects are the act-character, the *sense* or meaning, and the objectivity that is referred to. *Noetik* is concerned with the first, formal logic with the second, and real logic with the third. We are now told by Husserl that it is *Erkenntniskritik* or *philosophical logic* which shall complete the discipline of pure logic. *Erkenntniskritik* has to finally clarify the essence of theoretical reason. However, Husserl warns us, philosophical logic is not to relate directly to the various empirical sciences or empirical cognitions. It shall directly relate to the three other branches of logic, e.g., *Noetik*, formal logic and real logic, and only through, or by way of, *critiquing* these three and clarifying their fundamental assumptions, concepts and principles, it shall also, in the only possible way, provide the needed foundation for empirical cognition.

But what can philosophical logic add, e.g., to formal logic? Husserl throws light on this in a discussion of 'mathematical logic and philosophical logic'. He defends mathematical logic against distinguished critics such as Lotze and Windelband who simply lacked a deeper understanding of the nature of mathematical method. Nevertheless, the defenders of mathematical logic — in Husserl's view — are no less unclear about the epistemic value and the sense of the basic concepts and principles of mathematical logic. Mathematical method is symbolic and computational (*symbolisch-rechnerische*). The mathematicians are the technicians of deductive theory. The task of philosophical logic, however, is critical grounding and final evaluation. The two sorts of work need very different interests and abilities. The mathematician, Husserl adds, is not a pure theoretician, he does not pursue theoretical interest in the ultimate sense. He constructs technical works, systems, theories without insight into the essence of theory in general, and into the nature of concepts and laws that constitute it. The mathematician attends to what is for him a pregiven domain of objects, he has no sense for the subjective sources of the constitution of such objects. Philosophical logic needs to take up these problems that include: problem of givenness, problem of the relation between objectivity, ideality and subjectivity, problems, in the long run, of the very possibility of mathematical knowledge, of pure logical or formal-ontological cognition.

In a note from 1908, Husserl distinguishes between logical clarification and epistemological (*erkenntnistheoretischer*) clarification. An objective science aims at logical perfection — which means, not only full knowledge of its field, but a theoretical completion in the sense of bringing its objects to evident givenness and making its method match requirements for such evidence. This would mean, e.g., in mathematics, supplementing the symbolic-computational method by a 'constructive procedure' whereby 'mathematical objects' come to 'evident givenness'. Epistemological clarification would focus upon the deeper issue of the relation between a cognitive shape and the objectivities presented in it, as also between a sense and the object meant through that sense.

It is clear by now that what Husserl in these lectures of 1906-7 calls 'epistemological critique' or 'philosophical logic' appears later as his transcendental logic. In these lectures, the only reference, to 'transcendental logic' is to Kant's.

III

The discussion of 'form' and 'matter', the idea of the ultimate matter or stuff as the real individual, and the resulting thought that formal logic points towards such individuals as the ultimate values of its name and predicate variables, are all carried over into *Formal and Transcendental Logic*, esp. to the Appendix I. We have here a theory of hierarchy of form and matter which found its most well-known advocate in Heidelberger neo-Kantian Emil Lask who provided the most significant point of contact between neo-Kantianism and Husserlian thinking about logic. I will not pursue a comparison between the two theories of 'form' and 'matter' — Husserl's and Lask's — on this occasion, but will prefer to use the time left at my disposal to reflect on the thesis of the world-relatedness of formal logic and on the connected notion of a 'logic of truth' — both presented, in a developed form, in *Formal and Transcendental Logic*.

Kant seems to have made one of the distinctions between formal logic and transcendental logic to consist in this that transcendental logic considers thought as it relates *a priori* to objects while formal logic considers our cognitions in relation to other cognitions (*Kr.d.r.V,* A 57) simply with respect to the logical form. In so far as the object is constituted in accordance with the concepts and principles of the faculty of understanding, it is, in principle, true: transcendental logic is therefore, also called a logic of *truth* (A 62). Since Kantian critique presupposes the distinction between sensibility and understanding, and logic is concerned with the rules of understanding (*Verstandesregeln*), transcendental aesthetics falls entirely outside of transcendental logic, even if transcendental logic considers the applicability of the pure concepts of the understanding to the pure intuitions of space and time. I will comment on, how Husserl's transcendental logic (as well as Husserl's conception of formal logic) relates to these three Kantian claims: relatedness *a priori* to the world, the idea of a logic of truth, and exclusion of transcendental aesthetics from the purview of logic.

First, as to the world-relatedness of logic. According to Husserl — and it can be argued even according to Kant, despite the way he distinguished formal from transcendental logic — formal logic is no less object-constituting than transcendental logic. In his logic lectures (edited by Jäsche) Kant writes *'die allgemeine /= formale/ Logik auf alle Gegenstände überhaupt geht.'* (*Kant Werke* IX, p. 15). This precisely is Husserl's position. Kantian transcendental logic, the doctrine of the categories and the proof of their applicability to objects, is very much like what Husserl calls the *a priori* real logic — its transcendental feature being the thesis of the *constitution* of objects by the concepts of pure understanding. Husserl's transcendental logic does not take up the most general *material* concepts that hold good of the real world, but traces the origin of the formal logical concepts and laws to appropriate intentional acts in the subjective life of the transcendental ego. If formal logic, and so formal ontology, determines the forms of any objects whatsoever and therefore also of the most basic reals i.e., individuals, one wonders how Husserl could at the same time say that formal logical operations *refer back* to a world of individuals and so presuppose such a world? To this the proper answer, I think, would be: the two statements really amount to the same. If formal logic determines the forms of the individuals, it certainly presupposes the individuals whose forms it determines.

But to say that formal logic has an *a priori* relatedness to world is *not* to say that formal logic relates to *this* actual world. It rather relates to any possible world, and any possible world should consist in the long run of individuals *of some sort or other*. Keeping this distinction between 'this actually real world' and 'any possible world of individuals' in mind, we can ask: in what sense are we to understand Husserl's later thesis (propounded in *Formal and Transcendental Logic*) that the highest stratum of formal logic— built upon pure logical grammar and a pure logic of non-contradiction— is to be a logic of truth. We of course do understand that in the first two strata the semantic predicates 'true' and 'false' do not figure. We learn, e.g., from a note by Oskar Becker, how in the logic of non-contradiction the logical laws can be formulated without introducing the predicates 'true' and 'false', and how in the logic of truth the same laws have to be reformulated using those predicates. This is all fine. But we want to know more about the logic of truth.

Again, to return to Kant. When Kant characterizes transcendental analytic as a logic of *truth* in so far as the categories are shown to be 'objectively valid', it is a misleading characterization, for the concept of objective validity does not coincide with 'truth'. What is objectively valid is either true or false. Even the false cognition conforms to the categorial constraints. Kant concedes that logic cannot give a sufficient criterion of truth. Is a logic of truth in the strict sense possible?

Husserl's logic of truth may mean (a) theory of evidence, since he does take the true and the evident to be identical; or (b) taking the logic of noncontradiction to be an uninterpreted system, one may construe truth logic to be a model of that system in which the variables of that system are assigned objects (individuals and properties) in the world, so that to each model there corresponds one possible world, or (c) a theory about the essential *material* structure of the real world taken in its most general features. The last would include (α) a theory of basic individuals, properties and relations, (β) a theory of space and time, (γ) a theory about the material compatibilities amongst material contents or *core stuff*, and, finally, (δ) a theory of a harmonious unity of possible experience. This last understanding of 'Truth Logic' entails that the Kantian transcendental Aesthetic has to be a part of transcendental Analytic. In fact, what Husserl in the 1906/7 lectures calls '*a priori* real logic' or 'metaphysical ontology' becomes in the 'Conclusion' of *Formal and Transcendental Logic world-logic*, 'mundane ontology' (which explicates 'the all-embracing *a priori* of the eidos *world*), of which again, we are now told, transcendental Aesthetics forms the ground level.

Husserl's truth-logic remains an ambiguous notion hovering between being a stratum of formal logic (and then, a model theory with semantic predicates) and being a Kantian-sort of theory of any possible world of individuals (and then, includes an *a priori* theory of space and time and of the unity of sensuous experience) — an unstable middleground between a formal theory and a *material* theory. The crucial question remains: is an actual individual, quâ an actual individual, a possible object for logic? Further, what is the guarantee that there is a unique way of determining what the individuals constituting the *actual* world are? As long as we are, or Husserl is, unable to say that, the bare thesis that formal logic refers back to individuals is a formal, semantic thesis, not a 'material'/metaphysical thesis.

4. IN SEARCH OF THE ACTUAL HISTORICAL FREGE*

Hans Sluga has recently proposed reading Frege from an historical standpoint. He and several other Frege commentators have insisted that Frege's concept of sense is a cognitive and not a linguistic concept. With both of these suggestions, I am in general agreement. I will however comment on both and indicate how they are to be understood and within what limits.

I. Sluga rightly emphasizes the need for reading a philosopher from an historical perspective and in doing so provides a much-needed counterbalance for the a-historical and logical analytical approach to the writings of philosophers that one-sidedly predominates today. Because of this latter sort of approach, some philosophers have failed, according to Sluga, "to come to grips with the actual, historical Frege."[1]

Contrast this with Michael Dummett's views as to how one should read a philosopher. I am quoting from Dummett simply because Dummett is, at this point, responding to Sluga. "In interpreting a philosopher," Dummett writes, "there can be no substitute for thinking through, rigorously and in detail, what his arguments are and how they are supposed to work, what hidden assumptions must hold good if they are to be cogent, what answers could be given to objections, what relation one thesis has to another, in short, for subjecting his work to logical analysis."[2]

Are these — Dummett's and Sluga's — opposed points? I want to emphasize that they are not. That they are rather mutually complementary perspectives may be shown by the following two considerations. In the first place, Dummett's program of subjecting a philosopher's texts to logical analysis can be carried out only when we have been able to identify the various theses the philosopher holds. One obvious and indispensable way of doing the latter is by careful immanent textual exegesis. This by itself, one may contend, does not fully disambiguate the appropriate sentences and/or terms by eliminating all likely interpretations save the one that we want to assign to them. It is here that a historical perspective comes in handy. If we can determine which authors our

* Originally appeared in *Journal of the History of Philosophy* XXIII, 1985, 237-248.

philosopher read, whom he took more seriously than others he knew, or who most "influenced" him we can better carry out textual exegesis, and are more likely to be able to arrive at a definitive interpretation of the crucial terms, sentences, and arguments. You cannot fully understand, e.g., Kant's first two arguments on space in the Transcendental Aesthetic unless you determine whom he is arguing against. Dummett's program requires, in short, that we already ascribe to a philosopher a set of theses α, β, and γ. Once we have succeeded in doing so, we can ask what relation one thesis has to another, we can explore their hidden assumptions and test the cogency and consistency of the arguments. But whence the certainty that the theses are precisely those, namely, α, β, and γ, and not $\dot{\alpha}$ $\dot{\beta}$, and $\dot{\gamma}$? Both textual exegesis and historical reading can be beautifully combined to help us here.

In the second place, every philosopher is explicitly concerned with questions and problems which themselves can be asked only within a larger philosophical horizon. That horizon itself is not thematized within the philosopher's texts, but rather remains the unthematic but enriching background which first renders the concerns and questionings of the text possible. For the interpreter of the text, it is essential — if he is to grasp those concerns and questionings in their genuine import — that that horizon be rescued from anonymity. It is here that understanding the historical milieu within which an author thought is of the utmost importance. Logical analysis can fruitfully do its job only when this historical reading has been done. Neither can replace the other.

It would be an utter misunderstanding of the above claims on behalf of the historical reading if one took it to mean that what the historical perspective can bring out are answers to biographical curiosities such as why did Frege ask the questions he did? If this were the case, then, to be sure, the analytic project could make no important use of it and need not wait until the historical understanding is achieved. What the historical reading helps us determine are rather the *meanings* of the terms employed by philosophers, of the statements made by him and of the *arguments* advanced.

I think Dummett overstates his case when he takes Frege's saying that "for the logical concept there is no development, no history" (which Sluga quotes) to imply that the anti-historical approach of the analytic philosophers is the right approach to Frege. There are two confusions here: for one thing, it is not the case that an historical reading is appropriate only for an historicist philosopher but inappropriate for one who has no sympathy for the historicist mode of thinking. For another, to say that a philosophical text (or a philosopher) should be subjected to logical-analytical scrutiny amounts neither to saying that the text (or the philosopher concerned) or even what is worthwhile in it (or in his philosophy) is itself logical and analytical, nor to saying that the right sort of approach is that of contemporary analytic philosophy. Heidegger, no less than Frege, needs to be subjected to analytical scrutiny — but that neither demands recognizing that Heidegger's philosophy is logical-analytical nor amounts to making of him a contemporary analytic philosopher.

Note that I am distinguishing between that logical-analytical thinking which always characterized the best of philosophers in all ages and that narrowly defined — methodologically and contentually — philosophical school which goes by the name of 'analytic philosophy'. That concerns the second of the two confusions noted at the beginning of this paragraph. I will briefly explain what I mean by the way I have formulated the first confusion. Some philosophers, such as Hegel or Marx or Gadamer, are clearly historicists (admittedly, of different degrees and sorts); some philosophers, such as Plato, Kant and Frege, are non-historicists (again, admittedly of different sorts). It would be wrong to suppose that only the historicist philosophers deserve, and that their philosophies permit, historical reading, while the non-historicist philosophers do not call for, and their doctrines (such as is expressed by Frege's statement that for the logical concept there is no development, no history) rule out an historical and analytical reading. The thesis that a logical concept has no history can be illuminated by the historical perspective and scrutinized by analytical reflection, just as neither Marx nor Gadamer can rule out, by virtue of the specific historicist theses they may hold, the demand that their texts be subjected to analytic criticism.

The real issue, then, is not whether Frege's texts call for an historical reading or for an analytical reading. The real issue is: is it fair to look upon Frege as an analytic philosopher, or as the precursor of what is known as analytic philosophy? The fact that he is a founder of modern logic, the author of *Begriffsschrift*, or of logicism in philosophy of mathematics, should not be taken as entailing an affirmative answer. Perhaps the crucial question is, was he a philosopher of language, and if so, in what sense?

II. Those who understand Frege as an analytic philosopher or as the precursor of analytic philosophy agree, at the least, in ascribing to him a theory of meaning. Those who like Sluga challenge that interpretation do not want to read into Frege a theory of meaning. Sluga's formulation is cautions: "The Fregean notion of sense has its roots not in the theory of meaning, but in epistemology."[3] So also is Currie's: "... it is, I believe, quite wrong to say, as Dummett does, that Frege put the theory of meaning at the center of philosophy....If there was a revolution in philosophy which put the notion of meaning at the center, it came later with Wittgenstein, the Vienna Circle, and perhaps also with Tarski."[4] Both, I think, recognize that there are in Frege rudiments of what is called a theory of meaning; but, first, this theory was not the center of his philosophy, and second, even this theory had its origin in his deep epistemological concerns about the objective validity of mathematical knowledge. A pure theory of meaning should be a semantic theory; Frege's is a theory of cognitive contents. In fact, Sluga agrees with Tyler Burge that Fregean senses are cognitive contents, but does not quite agree with him that Frege never intended to give a semantic theory at all.[5] If I understand him aright, Sluga finds in Frege an unfortunate "conflation of semantics with epistemology."[6]

It appears to me that both groups of interpreters — those who make of Frege a theoretician of meaning, and so an analytical philosopher, and those who oppose that reading — share a certain understanding of 'theory of meaning', meaning is linguistic meaning, a theory of which should form part of a philosophy of language. To be able to ascribe such a theory to Frege would go a long way towards establishing him as an analytic philosopher; to be able to show, on the contrary, that Fregean senses are not linguistic meanings and consequently that Frege had no theory of meaning as such, would be to challenge that claim most effectively. I want to call this understanding of 'theory of meaning' into question, on the ground that although the dominant theories of meaning at present happen to be theories of linguistic meaning and also in the analytic tradition, there are also various other possibilities. One such is the Husserlian theory that an intentional act has its own meaning. Such a theory of act — or, let us say, cognitive — meaning would be a theory of meaning, and would have a bearing on the question of understanding of language as well, and yet cannot be accommodated within the confines of the analytic school. What, in effect, I am suggesting is that Michael Dummett's demand that a theory of meaning must be a theory of what we know when we understand a language, and the other camp's insistence — which in my opinion is the correct view on the matter — that Fregean senses are cognitive contents rather than linguistic meanings, are not only not incompatible with each other but need to be brought together to yield all the benefits of Fregean insights.

Let me begin by referring to Tyler Burge's thesis that Fregean senses are cognitive contents, not linguistic meanings.[7] This is a thesis with which I fully concur. But one wants to know what Burge means by 'cognitive contents'.

Fregean senses are to be taken as *cognitive* rather than linguistic, for at least the reason that they are introduced to explain the possibility of informative identity statements, to account for the difference in *cognitive value* between A = B (when true) and A = A. But what could be meant by saying that the senses are cognitive *contents*? For Burge, their cognitive character is brought out by the consideration that they are context-dependent whereas linguistic meanings are not. But this is not to say that all Fregean senses are context-dependent or that there are no context-dependent linguistic meanings. The two important aspects of Frege's theory which Burge draws attention to in support of his claim are: senses contain the "modes of givenness," and furthermore, they are components of abstract thoughts (or, thought contents). Also they are not linguistic meanings for the simple reason that they are eternal and objective structures, "ontologically and conceptually independent of language and of human agents."[8] In this account, there are several elements, not all of which seem to be mutually compatible. These are: (a) the *ontological* thesis that senses are abstract entities, timeless and objective; (b) the *cognitive* thesis that the senses contain modes of givenness; (c) the claim that they are cognitive contents; and (d) Frege's strong *anti-psychologistic* thesis. (b) and (c) are of a piece, but how are they, especially (c), compatible with (a) and (d)? How can we have a theory of cognitive *content* for which those contents are both ontological structures that are grasped and also ways of givenness of entities in the world? Obviously, we need a notion of cognitive content or conceptual representation that is non-psychological, as Frege's has to be. To say that it is either a thought or its component is not to say much, for one can come up with an account of 'thought' in terms of 'linguistic meaning'.

Looking for an account of that cognitive notion in Sluga's book, we are no less disappointed. What he does tell us is that the notion of sense is introduced to solve an epistemological problem. He also suggests that Frege's distinction between sense and reference has its origin in the Kantian distinction between concepts and objects. However, if there is any part of Frege's theory that shows the influence of Kant's theory of concepts (that concepts are predicative in nature, and that concepts without intuitions are empty), it is his theory of concept (that concepts are predicative and are unsaturated, to be completed by an object). Senses, being objects, are neither predicative nor need completion. The analogy between the Kantian distinction and the Fregean, I suspect, works somewhat like this: According to Kant, to know an object is to bring it under concepts. It is through concepts that we know whatever we know; there is no non-conceptual knowledge. Likewise, according to Frege, reference to objects is made possible through the media of the senses; there is no reference save through such mediation. But here the analogy ends. Concepts together with intuitions *constitute* the object; the senses do not constitute the referent. The disanalogy extends further than this, but I need not, for my present purpose, pursue it any further. The decisive question, in the present context, is this: can we say that senses stand in any relation to the human mind that is analogous to the way Kantian concepts do? Because senses are cognitive *contents* one would expect them to have some such relation to the mind, but the Fregean rejection of psychologism prevents us from ascribing to him such a view.

Curiously enough, Dummett also agrees that the notion of sense is a cognitive one. He also sees that for Frege, the notion of sense is connected with that of knowledge; sense is that which we know, when we understand an expression.[9] Yet, it is obvious that for Dummett sense is cognitive only because it is what we may be said to *know* when we understand a word or a sentence. To say that it is a cognitive *content* is tantamount to saying that it is an object of cognition. The idea of 'cognitive content' then loses any interesting significance.

How does Dummett continue, then, to regard Frege's philosophy to be a linguistic one? The crucial question, for him, is: How does one come to associate a sense with a word or a sentence? In what does our attaching a sense to a word consist? The connection cannot be inner mental connection, that would not only be inconsistent with Frege's anti-psychologism, but would make it impossible for the hearer to be sure that he associates the same sense as the speaker does. Consequently, Frege has to say that the sense which the speaker attaches to a sentence must be ascertainable from his observable linguistic behavior. Frege then must be a linguistic philosopher. *Our* only access to the sense is through linguistic expressions.

Surely, that the senses must be ascertainable from the speaker's linguistic (and non-linguistic) behavior is not explicitly said by Frege. Furthermore, if the original question, in what does our attaching a sense to a word consist, is answered by saying "through appropriate, publicly observable behavior," one may very well ask, "How does a use or behavior manifest a certain sense?" Or, perhaps, "How does one come to attach a sense to a certain use?" One needs an account that is other than an associationistic psychologism and a Wittgensteinian linguistic behaviorism. For this, again, we have to turn to a cognitivist reading.

The problem, then, for any Frege-exegesis which wants to look upon senses as cognitive contents is: How to reconcile this reading with (1) the ontological thesis that they are timeless, objective structures, (2) the complete rejection of psychologism, and (3) also the further fact that the theory of sense is also a theory of meaning. There is only one other theory that is available to us — and that too from amongst Frege's contemporaries, from one whom Frege knew — that suggests a way of reconciling these three claims. The theory is Husserl's.

III. The theory, in its broad outlines, goes somewhat like this. All intentional acts (in the Brentano sense) refer to their objects through their respective senses (or *noemata*). An intentional act (or state of consciousness) has thus a two-tiered structure. It is an event, occurring in someone's mental life, and is thus individuated by its ownership and its temporal location. It is a particular; it can itself be referred to, but cannot be "expressed" or communicated, and in this sense is psychologically subjective. At the same time, every such intentional act has a correlative sense or noema, its "content," which "belongs" to it, not as a real component, but as an "intentional correlate" which it must have, no matter whether the object to which the act refers exists or not. This sense or noema is not individuated by ownership or temporal location of the act whose sense or noema it is. In other words, one and the same noema can be the correlate of numerically distinct acts and so of acts occurring at different times and having different owners. Thus the noema or sense is intersubjectively available; in that weak sense it is objective; not being temporally individuated, it is "time-less." As determining reference, it also "contains" the mode of givenness; and being an intensional entity, it satisfies the cognitive criteria recognized by Frege. This theory is *not* the same as the theory of linguistic meaning, inasmuch as not all intentional acts are as such linguistic acts. However, the theory of noema is capable of generating a theory of linguistic meaning inasmuch as (a) linguistic acts, so-called speech acts, (which, and not the signs themselves, are the primary bearers of meaning on the theory) are a species of intentional acts, and so have their noemata which are linguistic meanings; and (b) all noemata, not excluding those of non-linguistic acts, are, in principle, linguistically "expressible" (though with suitable "adaptations"). The core of the thesis, which lends it its explanatory power, is the idea of intentionality as an act-noema structure, the idea namely that mental acts, although private particulars, have their correlative noemata which are publicly available, timeless, objective entities. It is through such noemata — as is the case in Frege's theory — that acts intend their objects.

IV. My purpose in introducing Husserl's theory in its bare outlines is not to suggest, even remotely, that Frege held some such view. It is also not intended to add to the historical story which Sluga has so admirably narrated, although there is no doubt, as Sluga recognizes, that Husserl (and the Brentano School in general) belongs to that intellectual milieu which must have shaped Frege's thought. (Note, for example, that the Brentano School, especially Carl Stumpf and Husserl, made a distinction between dependent and independent, complete and incomplete, contents, with which Frege's like-sounding distinction was closely connected.) The purpose of this digression into Husserlian phenomenology is to suggest one way of bringing the different, jarring components of Frege's theory into harmony.

Frege, to be sure, left these components at odds. It is because of this feature of Frege's thought that a realistic interpretation, such as Dummett's, appears so convincing if one considers large chunks of what Frege says sundered from his overall concerns, from his historical rootedness as well as from some of his enigmatic pronouncements. And yet, when one takes all these latter considerations seriously, the picture emerges of what Sluga calls an objective idealist. Objective idealism, it should be borne in mind, makes room for a limited realist discourse.

But 'objective idealism' itself is a blanket title. It covers a large variety of philosophical positions. It is nevertheless helpful to recall that objective idealism of any sort must be committed to (a) a conception of consciousness that is non-psychologistic and (b) a conception of "contents" of consciousness that, like consciousness itself, is not a private particular. Frege's theory of sense, interpreted as cognitive content, satisfies the second requirement, but can we ascribe to him anything even remotely suggesting (a)? One who wants to interpret Frege as an objective idealist can turn to enigmatic texts such as these three:

A. "By objectivity I mean independence from our sensing, intuiting, and imagining and from the construction of internal pictures from memories of earlier sensations, but not independence from Reason..."[10]

B. "Neither logic nor mathematics has the task of investigating souls and the content of consciousness whose bearer is an individual human being. Rather, one might perhaps set down as their task the investigation of the Mind — of the Mind, not of minds."[11]

C. "But still the grasping of this law [the law of Gravitation] is a mental process! Yes, indeed, but it is a process which takes place on the very confines of the mental and which for that reason cannot be completely understood from a purely psychological standpoint."[12] [In a footnote to C, Frege adds:

> "I should say that this question of grasping thoughts and recognizing them to be true is still far from being grasped in all its difficulty."]

Although Frege's rather naive philosophy of mind remains in sharp contrast with his remarkably sophisticated logical theory, these three passages show that he felt the need for a richer philosophy of mind which could support his theories. I have characterized his "official" philosophy of mind as "naive," for he looked upon mental life as consisting of private particulars, and the science of it as unavoidably inductive and statistical. It is this conception of the mental (which is equated with the psychological) that provides the underpinning for both his and Husserl's anti-psychologism. That the mental in this sense has to be sharply sundered from the logical, is obvious. The metaphor of 'grasping' (the subjective mental act grasps an objective sense), which lends support to a realistic ontology of the senses, certainly provides the initial and, as a first approximation, the correct description. But the problematic character of this metaphor of 'grasping', as emphasized by passage C, shows that Frege was groping not only for a more suitable metaphor, but also for a more satisfactory concept of the mental act and its relation to the thought or sense that the act is said to grasp. But *did* he arrive at any such account? In order to be able to do so, he needs to satisfy (a): He needs to have a non-psychologistic theory of consciousness. Passages A and B suggest *some such* concept, but the suggestions are too vague to admit of a precise interpretation.

V. An important part of the argument to the effect that the Fregean senses are not linguistic meanings but cognitive contents rests on Frege's statements about indexicals.

I want to suggest that anyone who reads Frege from a cognitivist perspective, as I am doing along with Burge and Sluga, cannot but be interested in the significance of his remarks on indexicals and on proper names strictly so-called.

With regard to indexicals, Frege held — as is well-known — that the sense of 'I' varies from speaker to speaker. (He does not say that the sense of 'I' varies even when the same speaker utters it on different occasions.) Frege also writes that everyone is originally given to himself in a special and unique manner in which he cannot be given to anyone else. In so far as the sense contains the mode of givenness, the sense that goes with every speaker's utterance of 'I' is something unique and incommunicable. Frege is thus led to a theory of incommunicable sense which does indeed appear to conflict with the original theory of sense. It would have been a less complicated theory if only Frege had held that indexicals denote or refer and have 'linguistic meaning', but do not have any *sense*. But he could not have held such a view. If the sentence "I am in pain" is to express a thought, such thought can consist only of the senses of the component parts of the sentences. A thought cannot have as one or more of its components, a referent. Therefore, 'I' must express a sense.

Now how are we to understand this thesis about the unique, incommunicable *sense* of 'I'? Husserl, who *largely* shared Frege's views on such matters, wrote: "each man has his own I-presentation (and with it his own individual notion of I)."[13] Could one not retort that this individual I-presentation is a *Vorstellung* and not a *Sinn*? In order to be an incommunicable sense, it must, first of all, be a sense, and not a *Vorstellung*. If the I-presentation consists in a set of descriptive predicates that I alone believe to be true of myself, then there is no reason why I cannot let others know what I *take myself* to be: Such a sense would not be in principle incommunicable. We have then to look for the incommunicable sense somewhere in between these two extremes: the availability of a descriptive account (in which case we have a Fregean sense but no incommunicability) and the varying modes of self-*feeling* (in which case we have incommunicability but not a sense). Could it consist in an *unvarying* mode of self-*understanding*, a non-conceptual, inarticulate self-*interpretation* in terms of (Heideggerian) projects and choices? I do not know, but we are bordering on the concept of mode of self-givenness on my own part that is neither a set of descriptions nor a passing self-feeling — the idea of *pre-conceptual sense*.

We may be better off with regard to proper names. Let me take courage to return to that much-discussed footnote in "Sinn and Bedeutung" in which Frege writes that in the case of an actual proper name such as "Aristotle" opinions as to its sense may differ. It does not matter if Frege would have wanted, as he certainly appears, to eliminate such variations in sense in an ideal scientific language. Sluga has most appropriately shown that in the *Begriffsschrift* account 'the mode of determination' is a property of the sign, whereas in the account of 1892 it is contained in the sense of the sign. One may then reasonably maintain that a proper name such as 'Aristotle' itself is not associated with one unambiguous sense and with no other, that it does not make sense to say that of the many senses different persons may attach to it, there is a *true* sense. One may then hold that since the sense after all contains the mode of presentation, what sense one attaches to the proper name depends upon how the person Aristotle is presented to one, the speaker or the hearer. After all, Aristotle may be presented, thought of, referred to in different manners, from different perspectives. Senses do not originally belong to signs. Signs derive their senses from the interpretive acts of the speaker or hearer or the community. To ask then what is *the sense* of 'Aristotle' would appear to be misconceived. To ask, what is the *essence* of Aristotle — a different question altogether — makes sense. The sense of 'A' and the essence of A need not coincide.

Thus although Frege began by attaching senses to signs, with the increasing emphasis on the idea of mode of presentation, the senses come to be tied closer to what could be called intentional acts. And yet Frege did not come to develop a theory of intentionality and a philosophy of mind adequate for this purpose.

These are two directions to move ahead from here on. Either one may want to go all the way with the implied cognitivism, which I prefer, or one may want to return to the purely semantic theory of the *Begriffsschrift* (which, I presume, Sluga prefers).

All of this bears on the issue of psychologism. Why is it that Husserl in his maturer years conceded that although his *refutations* of psychologism in the *Prolegomena* of 1900 were still valid, he had not then grasped the true nature of psychologism and so had not *overcome* it? Germane to psychologism is a naturalistic interpretation — and here Sluga's historical reading is, in my view, accurate — not alone of the logico-mathematical entities, but of the mental life itself. The mental as such is *not* the psychological. It *becomes* the subject-matter of psychology as a natural science by being, in that very thematization, inserted into the causal order of nature. What psychologism and anti-psychologism of the common variety *both* share is a conception of the mental that is tied to a certain conception of psychological discourse. It is not necessary to deny that such discourse is legitimate or even to say that the mental is not a private particular. What I want to insist upon is that what makes psychologism possible in the first place is that interpretive framework within which the life of consciousness comes to be construed as the domain of private particulars or as the subject matter of introspective or even of scientific psychology. One may be said to have truly overcome psychologism when one recognizes that we have here, underlying psychologism, an interpretive framework that remains hidden from view, and so pretends to be an ontological thesis. One then may be led to ask *how* this framework is possible. An unavoidable precondition for raising this last question is to subject that framework to the famed 'epoche'. We are then on the threshold of a transcendental philosophy.

The obvious presupposition of such a *radical critique* of psychologism is a conception of consciousness as *transcendental.*

What, then, was the actual historical Frege's philosophy like? Was he a realist or an idealist? Was the core of his problematic semantics or epistemology? I think any interpretive choice here is to be dictated as much by Frege's actual historical texts as by the interpreter's 'prejudices'. The texts situate us in an ambiguous, rather unstable position in relation to various possibilities. In metaphysics, his thought is torn between a Platonic (and commonsensical) realism and a sort of Kantian idealism. In theory of meaning, it occupies an unstable position between a semantic theory (which attaches senses and references to signs) and a semi-contextual, *cognitivist* and epistemological theory which makes senses and references functions of contexts, intentions, and cognitive perspectives. His philosophy of mind contains the glimpses of the concept of intentionality, but does not go beyond a logic of intensionality. The mental act is relegated to empirical psychology, and yet the 'grasping' of thought, as a mental act, calls into question the gulf that divides the mental and the logical.

The actual, historical Frege is elusive. We, the interpreters, make of him what we want by putting pressure on one or other components of his thought. In doing this, we shall be carrying *forward* the historical relevance of Frege, even if our reading is not as meticulously historical as Sluga's. In this regard, the analytic interpretation over the last fifty years has let Frege as it saw him create history. Can the new Frege that is emerging — the cognitivist, epistemological, phenomenological Frege — equally have his day?

NOTES

[1] Hans Sluga, *Gottlob Frege* (London: Routlege and Kegan Paul, 1981), 3.

[2] Michael Dummett, *The Interpretation of Frege's Philosophy* (London: Duckworth, 1981), 528.

[3] Hans Sluga, *Gottlob Frege*, London: Routledge and Kegan Paul 1980.

[4] Gregory Currie, *Frege, An Introduction to His Philosophy* (New Jersey: Barnes and Noble, 1982), 167.

[5] Hans Sluga, "Semantic Content and Cognitive Sense," (unpublished paper), fn. 4.

[6] Hans Sluga, *Gottlob Frege*, 154.

[7] Tyler Burge, "Sinning against Frege," *Philosophical Review*, 88 (1979): 398-432.

[8] Ibid.: 405.

[9] See esp. Michael Dummett, *Frege: The Philosophy of Language* (London: Duckworth, 1973), 679.

[10] Gottlob Frege, *The Foundations of Arithmetic* (Oxford: Blackwell, 1953), 36.

[11] Gottlob Frege, *Kleine Schriften* (Darmstadt: Wissenschaftliche Buchgesellschaft, 1967), 359.

[12] Gottlob Frege, *Posthumous Writings* (Oxford: Blackwell, 1980), 145.

[13] Edmund Husserl, *Logical Investigations*, trans. J.N. Findlay (New York: Humanities Press, 1970), 1:316.

5. DUMMETT, FREGE, AND PHENOMENOLOGY*

In this companion volume to *Frege: Philosophy of Language*, Michael Dummett responds to some of the critics of that first book and also examines some major attempts at interpreting Frege. In doing all this, he, as in the first book, takes up and discusses various hotly debated issues in contemporary philosophy of logic, semantics and epistemology. There may be disagreement as to how successful he is in throwing new light on these issues, but Dummett, to my mind, clearly succeeds in establishing his credentials as the *doyen* of Frege scholars. He demonstrates superb mastery over Frege texts and skill in handling and interpreting them. Furthermore, in this book he achieves greater focus on the issues of Frege exegesis, something readers of the first book must have missed. This focus is mainly due to the context. Dummett is criticising other interpretations, so he quickly settles down to the texts and interpretive issues, although the tendency — agonisingly marked in the first book — to stray away from the issue under consideration is not altogether absent here too.

In this review, I will single out only a few of the questions, those which would be of special interest to readers of this journal. These are: (1) the principle of interpretation which Dummett defends, (2) Frege's relationship to Husserl, (3) Frege's theory of *sense*, (4) the context principle, and (5) the issue about realism vs. idealism. I will preface my discussion of Dummett's views on these questions by quoting a remark on Frege. Heidegger wrote in 1913:

> G. Freges logisch-mathematische Forschungen sind meines Erachtens in ihrer wahren Bedeutung noch nicht gewürdigt, geschweige denn ausgeschöpft. Was er in seinen Arbeiten über "Sinn und Bedeutung", über "Begriff und Gegenstand" niedergelegt hat, darf keine Philosophie der Mathematik übersehen; es ist aber auch in gleichen Masse Wertvoll für eine allgemeine Theorie des Begriffs.[2]

Unfortunately, many phenomenologists, no matter of which camp, do not recall this. Reading Frege can only help the phenomenologists to understand themselves better and with greater clarity.

* Originally appeared in *Journal of the British Society for Phenomenology*, Vol. 15 No. 1, January 1984.

I

Dummett attacks the so-called historical method of interpretation, defended and used by Sluga in his recent book on Frege. I am not quite sure of exactly what Sluga means by the historical, as opposed to the analytical, method of understanding and interpreting a philosopher. So whether Dummett is fair to Sluga, I will not judge. What Dummett attacks however is the contention that a philosopher is best understood by determining the historical influences that shaped his views and also by bringing out parallels to his ideas in the works of his predecessors. The "talk of influences and tendencies", Dummett remarks (p. 529), characterises art history rather than philosophy. Dummett's own view, on this matter, is best stated in the following sentence:

> In interpreting a philosopher, there can be no substitute for thinking through, rigorously and in detail, what his arguments are and how they are supposed to work, what hidden assumptions must hold if they are to be cogent, what answers could be given to objections, what relation one thesis has to another, in short, for subjecting his work to logical analysis (p. 528).

To be fair to Dummett, it must be mentioned that he does recognise the need for historical study. Historical research can help us, he concedes, in several ways. By knowing well the epoch to which an author belonged, one can avoid making use of things that were not yet known when he was writing and one will be able to take account of the views the author thought himself to be opposing (pp. XVI, 528). By knowing the views the author held in common with his contemporaries, one will be able to better appreciate his original ideas as distinguished from those he derived from the age (p. 529). In the end, however, a philosopher's writings have to be read analytically in the manner stated by Dummett in the sentence quoted above.

While in general I agree that a philosopher, especially one like Frege, must be read "analytically," (using that modifier in a sense that has nothing to do with the name "analytical philosophy") as stated by Dummett, I must also point out that historical research is of paramount, and not merely ancillary importance. A brief explanation of what I mean may be given thus: if a philosopher asserts a set of propositions p, q and r, the interpreter, according to Dummett's principle, has to ask: what is the logical relation between these three propositions? Are they mutually consistent? If r is taken by the philosopher to follow from p and q, and if it appears it does not, should we ascribe to the author another "hidden" premise q^1 jointly with which p and q will logically "yield" *r*? Granted all this, we still have to interpret the sentences used to express, allegedly, the propositions *p, q* and *r*. And it is here that the historical approach would appear unavoidable. To interpret the meanings of an author, one needs to look into the historical context in which he thought and wrote. Once the meanings are interpreted and fixed, we will have at hand the set of propositions to be subjected to logical analysis. While, it seems to me, this is true in general of all authors, for philosophers like Frege whose concern was rather a-historical, the historical context will be narrowly limited, but nonetheless not altogether irrelevant for purposes of interpretation. A complete neglect of this context — and in this Sluga may be right — leads to the sort of misinterpretation of Frege which has prevailed in this country since Carnap, which misconstrues the Fregean 'Sinn' as 'linguistic meaning'.[3] At the other extreme may be Frege's alleged indebtedness to Lotze on which Dummett's book contains some valuable historical research.

II

The question of 'influence' belongs not merely to art history, but also to history of philosophy. And, it is not infrequently the case that wrong historical judgments about influences have either led to, or have been sustained by, wrong interpretations of the authors concerned. An interesting example is the relation between Frege and Husserl. Dummett appears to share in the widely held belief that it was Frege, especially by his sharp review of Husserl's *Philosophie der Arithmetik*, who led Husserl to reject the psychologism he had held in that early work and develop the conception of a *pure* logic. Since I have elsewhere argued against this view of the matter,[4] I will not take it up now. What I will take up here for consideration is the view that Husserl generalised Frege's distinction between sense and reference to mental acts. This interpretation has been proposed by Føllesdal, not so much as a historical judgment but as a principle which may be helpful in understanding Husserl, particularly when one is approaching him from the perspective of analytic philosophy. It is not quite clear in what manner Dummett wants to understand this thesis. Considered as a historical judgment, it is untenable. First of all, it is questionable if Husserl took over the distinction from Frege. But, more importantly, he was not generalising the *Fregean* distinction to all mental acts, but a distinction which, although seemingly Fregean, is not quite so. The Fregean distinction concerns signs, the distinction with which Husserl starts, before generalising it to all acts, already concerns a specific sort of *acts*, namely, the expressive or *ausdruckende* acts. If Husserl had started with a distinction between sense and reference with regard to signs, and *then* generalised it to all acts, his procedure would have been, as Dummett complains, "completely wrong-headed". *But from the beginning,* Husserl's semantics is a speech-act semantics, for him meanings are meanings of acts. He is concerned, even before the alleged "generalisation," with *la parole* and *not* with *la langue*. He is on the side of Searle and not on the side of Davidson. It is not surprising that Dummett finds here (i.e., in the alleged generalisation) the seed of Husserl's relapse into psychologism. To regard senses as in some manner mental contents rather than as objects independent of the mind to which the mind is related is, for Dummett, "to obliterate what was, to Frege, the crucial distinction between senses and ideas" (p. 58-9). In this criticism of the Husserlian move, Dummett is faithful to Frege. He shares with Frege the latter's naive and impoverished conception of the mental, according to which the mental = the private/non-communicable = the psychological. Given this conception of the mental, it is of course absurd that the sense or the noema, which is objective and communicable (leaving aside, for the present, Frege's thesis about "incommunicable senses"), would at the same time be regarded as "mental contents." However, one of Husserl's important philosophical insights is that although the senses could not be "ideas" in the sense of the British empiricists from Locke to Hume, and although logic and semantics have to be kept apart from empirical associationistic psychology of any sort, it by no means follows that the senses have nothing to do with the acts of consciousness. In fact, that the acts of consciousness are irreducibly mental, private and psychological is a consequence of a certain interpretive framework, which is what Husserl overcomes. On his theory, an act of consciousness is two-tiered: at one level, it is a private, temporally individuated event, at another level, it has a *structure* that is not private, not temporally individuated, that is therefore objective, i.e. intersubjectively available and communicable.[5] It is only this new concept of consciousness which makes it possible to hold *both* that the senses are non-psychological *and yet* are structures exemplified in acts of consciousness. Psychologism is thereby *truly overcome*, for one is rejecting that very interpretation of mental life which the naive anti-psychologistic thinker had shared in common with the naive psychologistic philosopher. You do *not* truly overcome psychologism by holding on to psychologism's underlying philosophy of mind while rejecting its reduction of senses and referents to the mental in his sense of that term. What one needs is a picture of the mental life that is richer than the empiricists'. Husserl provides us with one.

Husserl, then, is neither *generalising* a linguistic-semantical distinction to all mental acts, nor is he *applying* the Fregean distinction to certain mental acts, namely to those that involve propositional attitudes. He is rather bringing semantics, philosophy of language and philosophy of logic under a rich philosophy of mind, i.e. a theory of intentionality, steering clear of the Scylla of psychologism and the Charybdis of a naive Platonic realism.

III

Dummett shows with remarkable skill that Frege's notion of *sense* has several 'ingredients' which are not always in harmony with one another. These 'ingredients' are:
(a) sense as the mode of presentation,
(b) sense as conventional, linguistic meaning,
(c) sense as constituent of thoughts.

Evans takes (a) alone as the Fregean sense, and so concludes that if an expression lacks reference, it must also lack sense, for if nothing is presented, one cannot speak of a mode of presentation. Dummett rightly rejects this un-Fregean position. The second ingredient is important for Dummett's understanding of Frege; it is however something that determines reference without any dependence on the context, and yet Frege does say that in many cases we need to know the context in order to move on from sense to reference. For this latter sort of cases, the mode of presentation is the relevant ingredient. Coming now to the third ingredient, Dummett aptly distinguishes between two, both Fregean, senses of 'thought': thought as the bearer of absolute truth-values, and thought as the object (or content) of propositional attitudes. Dummett then shows how in the case of indexicals and in the case of *oratio obliqua*, there are tensions amongst these ingredients. In all this, I think, he succeeds remarkably well. I will restrict my comments to three: first, while recognising that the Fregean notion of sense has these different, often mutually jarring, ingredients, Dummett yet asserts: "That does not entail that it is mistaken to interpret Frege as virtually identifying sense with conventional significance, or, rather, with a particular aspect of it..." (p. 108). Dummett's point is that it is only in the case of token-reflexive expressions and *oratio obliqua* that the divergence amongst the various ingredients becomes noticeable; in all other cases, they so coalesce as to leave the linguistic meaning as the only relevant factor for consideration. I think in this inference, Dummett's interpretation is too hasty. Why not say that the Fregean sense always has these "ingredients," that it is (a) an objective entity which bears absolute truth values; (b) the content of mental acts, especially propositional attitudes; (c) contains the mode of presentation; and (d) can be linguistically expressed? With this, we are indeed closer to the Husserlian notion of the noema.[6]

Secondly, we are required to decide what is to be meant by Dummett's talk of "ingredients." Certainly, Dummett does not mean by saying that the Fregean notion of sense has many "ingredients," that there are many different senses in which Frege uses the term "Sinn." What he means is that these are different *components* or better still, *aspects* of the *Sinn*. What the locution would mean, then, is that the Fregean *Sinn* is at the same time a mode of presentation, an objective thought or its constituent, an object of a propositional attitude, and a linguistic meaning. How can it be all these at the same time? I think, trying to think this through will take us beyond the bounds of Fregean philosophy of mind and a merely linguistic theory of meaning.[7]

Finally, there is a tension that runs through Dummett's own reflections on Frege. On the one hand, he rightly emphasises that the Fregean *Sinn* is a cognitive notion, and "therefore represents the insertion of a cognitive element into the theory of meaning;" (p. 530) on the other hand, he struggles hard to maintain that Frege is a linguistic philosopher. Dummett detects in Frege a fundamental tension (which pretty well corresponds to the tension in Dummett's own reflections on Frege) between the claims to primacy of private understanding and of public meaning. While rightly emphasising that Frege cannot be taken as regarding public meaning as derivative from private understanding, Dummett yet concedes that "there is considerable reason to interpret him as holding just such a view of sense" (p. 113).

Phenomenology provides a viable way out of this dilemma by asking us to revise our concept of the mental. The mental is not wholly private: the occurrent act is, but its content or noema is not. The sense can very well be the content of private understanding and yet be expressible in public language. The empiricist idea that to be the content of a mental act is to be a private particular or something "psychologically real" (Evans), has to be rejected.

IV

Dummett's discussion of Frege's Context Principle (C.P.) is masterful. I will here briefly recall only some of his contentions, and at the end insist on their significance for phenomenology.

First, the Context Principle is different from the priority principle, the principle namely that we do not start from concepts and compound the judgments out of them, but rather start with judgments and decompose them into their parts. Thus whereas the priority principle applies to concepts and functions, the Context Principle applies to all words.

Secondly, the Context Principle is not incompatible, contrary to a common misconception, with the requirement Frege lays down, that words must have unique and invariant senses and referents in all contexts. As Dummett puts it, the Context Principle "is a principle concerning what it is for a word to have a meaning, and does not imply that its meaning may legitimately vary from one occurrence to another" (p. 364).

Thirdly, Dummett is able clearly to distinguish between the Context Principle as a principle about sense and the same as a principle about reference. With regard to sense, the principle says, according to Dummett, that the sense of a word consists in the contribution it makes to the senses of the sentences in which it does occur (pp. 369-380). Considered as a principle about reference, it says — according to Dummett — that:

> What justifies us in regarding an expression as standing for an object is, first, that it behaves like a proper name, and, secondly, that a definite sense has been provided for every sentence in which it may occur (p. 382).

Fourthly, how is the Context Principle to be reconciled with the composition principle, the principle, namely, that the sense (or, reference) of a sentence must be composed of the senses (or, references) of the expressions occurring in that sentence? Here Dummett's solution runs as follows: the sense of the sentence is primary in order of explanation (of what it is for a word to have a sense), while the sense of a word is primary in order of recognition (p. 547). If one does not understand a component word, one does not understand the sentences in which it occurs; at the same time to understand a word is to know what contribution it makes to the sense of the sentence in which it occurs.

Finally, if a word has a meaning only in the context of a sentence, could not one take a further step ahead and hold that a sentence has a sense only in the context of a paragraph. But why should one stop with a paragraph? One needs to expand the context to the entire language. One must, then, in some sense, understand an entire language in order to understand an expression belonging to it. Such a holism has something intuitively absurd about it. Dummett rightly emphasises that Frege's Context Principle "singles out the relevant segment of the total context as the *sentence*, and so gives sentences a special, distinguished role in language and hence in any theory of meaning" (p. 370). A sentence expresses a complete thought which can be grasped on its own. Recognition of the primacy of self-complete character of 'apophansis' or predicative judgment underlies the rise of western logic since Aristotle. Frege's Context Principle does not overstep the limits imposed by this tradition. In this, he and Husserl are on common ground, although Husserl was willing to go beyond the *naivete* which consists in focusing on the judgment as a fixed, autonomous entity and to look at the larger horizon of mental life from which it is carved out, against which it displays its figure, and within which it is constituted.

It should be of interest for the phenomenologists that an extended, liberalised Context Principle could provide an access to the holism of the hermeneutic philosophies. Frege and Husserl agree in rejecting epistemological atomism without bouncing to the other extreme of an unmitigated holism.

V

Dummett's views about 'realism' are well known and so I will not deal with them in this review. What I will do, on the contrary, is take into account a statement that Dummett makes about the conditions of 'an intellectualized form of idealism'. Dummett makes this statement in connection with his discussion of a proposal made by Evans and McDowell to emend Frege's theory of sense. Evans and McDowell want to preserve Frege's realism by excluding senses with no corresponding reference. The sense, for them, is only the mode of presentation, not an intermediary object which it would become should we admit senses without corresponding reference, and once we have such an intermediary object, there is no guarantee that we have ever anything beyond the domain of senses. It is presumably this last possibility — the view namely that the alleged object is *constituted* by senses — which Dummett calls the "intellectualized form of idealism."

This form of idealism, Dummett writes, is made plausible by a combination of the following three theses: "that we apprehend an object only as presented in some particular way; that the mode of presentation is a sense; and that to *no* sense is it intrinsic that there is a corresponding reference' (p. 138). Frege, according to Dummett, avoids idealism by rejecting the third thesis and making room for modes of presentation which guarantee reference. As regards abstract and logical objects, Frege did provide formal criteria for expressions to have guaranteed reference; as regards terms of concrete objects, as Dummett recognises, to determine how reference may be guaranteed, one needs an epistemological theory which Frege did not offer (p. 141).

Let me now return to Husserl. Husserl's view that the object is *constituted* by its noemata (taken apart from the further thesis that the constitution is eventually by *acts*) is a form of 'intellectualized idealism'. Of the three theses whose combination, according to Dummett, makes such idealism plausible, Husserl undoubtedly holds the first: namely, that we apprehend an object *only* as presented in some particular way. Husserl, along with Frege, also holds the second thesis, *not* exactly in the form Dummett states it but in the modified form that the mode of presentation is a component of the sense. Crucial for this discussion is the third thesis. Does Husserl hold the view that to *no* sense is it intrinsic that there is a corresponding reference? Before answering this question, however cursorily, I should point out that for the purpose of preserving realism both the extreme positions need to be avoided: the thesis that for all senses reference is guaranteed, and the thesis that such guarantee is not available for any sense. In the former case, reference would become a function of sense, the distinction between sense and reference would collapse in favor of sense; in the latter case, guaranteed reference *not being available at all*, the consequence would seem to be at best scepticism about reference, and at worst an idealism of a sort.

For understanding the nature of Husserl's idealism, we need to ascertain his likely answer to the question about the availability of guaranteed reference. I cannot, in this review, undertake a discussion of this issue. It may suffice to point out that whatever may be the right answer, two considerations must be taken into account in determining Husserl's answer: first, one justification of phenomenological reduction lies in the fact that the object of intended reference may *not* be there, so that neither the existence nor the nonexistence of the object necessarily follows from the structure of the noema; secondly, there is a type of sense called by Husserl "fulfilling sense" (in the case of perception, "perceptual noema") which presents the object as "bodily selfgiven" (*leibhaft gegeben*), and so may be regarded as guaranteeing reference.

This brief discussion of Dummett's book, I hope, would show that not only the writings of Frege, but also those of Dummett, should be of profound interest to Husserlian scholars.

NOTES

[1] Michael Dummett, *The Interpretation of Frege's Philosophy*, London, Duckworth, 1981.

[2] M. Heidegger, *Gesamtausgabe*, herausg. F.-W. von Herrman, Bd. 1, Frühe Schriften. Frankfurt: Vittorio Klostermann, 1978, p. 20.

[3] Cp. T. Burge, "Sinning against Frege," *Philosophical Review*, LXXXVII, 1979, pp. 398-432.

[4] J.N. Mohanty, "Husserl and Frege: A New Look at their Relationship," *Research in Husserl's Logical Investigations*, The Hague: Martinus Nijhoff, 1977.

[5] See A. Gurwitsch, "The Kantian and Husserlian Conceptions of Consciousness," in: *Studies in Phenomenology and Psychology*, Evanston: Northwestern University Press, 1966, pp. 148-174.

[6] For the structure of the *noema*, see Husserl, *Ideas* 1, Chapters 9, 10 and 11.

[7] I suggest this in my book *Husserl and Frege* (Bloomington: Indiana University Press, 1982).

6. HEIDEGGER ON LOGIC*

Why should one write on Heidegger's understanding of logic? After all, Heidegger was not a logician, nor did he do philosophy of logic. Indeed, there is no justification for expecting of any great philosopher whatsoever that he should have views, and reasonably plausible views, about the nature of logic or on specific themes belonging to the domain of logic. A moral philosopher may totally bypass any concern with logic, without detriment to his thinking. As an existentialist philosopher, Heidegger could have done that, and much of his *Dasein*-analytic would yet have retained its value. But Heidegger was also an ontologist, and was deeply concerned, all his philosophical career, with metaphysics and with the various questions about the nature of thought and being. These concerns, to say the least, bring him to the proximity of logic as it had been understood in the tradition going back to Aristotle. And, as a matter of fact, Heidegger's own access to the problems of ontology and metaphysics has been determined by his reflections on logic. Two claims may therefore be advanced. First, it is not unreasonable, and what is more important, not unfair to Heidegger, to enquire into his understanding of logic. Secondly, his reflections on logic may help us to gain a better understanding of his overall philosophical interests than would be possible otherwise. Even if he was not a logician he was concerned with the nature of logic, and with some central problems belonging to the domain of logic. This concern begins with his doctoral work on the problem of psychologism in the theory of judgment,[1] continues in the habilitation work on the semantic categories in Duns Scotus[2], and reaches its maturity in the Marburg lectures of 1925-28.[3]

In this essay, I will deal with three topics. In the first section, I will try to determine how Heidegger understood the nature of logic. In the second section, I will consider the one problem of logic to which he devoted a great deal of attention: the theory of judgment. In the third section, I will look into how this concern with logic opens up for him several paths to go beyond logic. At the end, I will reflect on this entire account, not so much to find faults with

* Originally appeared in the *Journal of the History of Philosophy*, Volume XXVI, Number 1, January, 1988, pp. 107-135.

Heidegger's understanding of logic, as to determine its precise nature and limitations.

1. NATURE OF LOGIC

A. A Preliminary Definition

One commonly held view of the nature of logic, in the traditional accounts, is that logic is a normative science of thought, whose aim is to lay down those *rules* which one *ought* to follow if one aims at truth. This account may be faulted on various grounds. First of all, 'thought' is ambiguous, referring both to the process of thinking and to the content of thinking. Of these two, the former belongs to the field of psychology. If the content of thinking is understood in the sense of objective meanings or structures of meaning, propositions or configurations of them, then only logic may be said to be concerned with them. Why then is logic to be still regarded as a normative science? Of course, once there is a logical law to the effect 'If p implies q, and p, then q' (where p and q are propositional variables), then it does follow that if a person believes in a proposition 'A implies B' and also believes that A (where 'A' and 'B' are the names of propositions), then he also ought to believe that B. But such a normative demand on the person's rationality is no part of the business of logic. Finally, the term 'truth' is ambiguous, referring both to material truth (the sense in which the statement 'it is raining now in Norman' is true if and only if it is in fact raining in Norman) and formal truth or validity (the sense in which the inference "All men are immortal, all Greeks are men, therefore, all Greeks are immortal" is valid, being a substitution instance of a logical law, even if one of its premises as well as its conclusion are materially false). It may appear as though logic is concerned with validity, rather than with truth understood, as it usually is, in the first of the two senses. If we accept these three emendations, then we can transform the initial account of logic into some such as this: logic is the science of meaning-structures in so far as they are valid. On this account, the task of logic is to lay down the laws of validity of meaning-structures.

Heidegger, under the influence of Husserl's idea of a pure logic of meaning, concludes his dissertation with a formulation of the task of logic that is very much like the one we have just arrived at. The logician, he concludes, must aim at bringing out the precise meanings of sentences and then proceed to determine the forms of judgments according to objective differences of meanings and their simple or compound structures, and bring such forms into a system.[4] Although the notion of validity does not figure in this account, the way forms of simple meanings and compound meanings can be brought into a system must be by showing the relations of implication amongst them, and the laws of their implication should be able to yield laws of validity of meaning-structures. But Heidegger has no doubt, in those early works, that the proper logical object is neither the mental process of thinking nor the reality (whether physical or metaphysical) about which one thinks, but the *Sinn*, understood both as the meaning of a sentence and as the identical content of judgment.

B. Critique of Psychologism

Such a preliminary account of logic already implies a rejection of psychologism. Heidegger is aware of Frege's rejection of psychologism, but it is Husserl, who, he writes, "has systematically and comprehensively laid bare the essence, the relativistic consequences and the theoretical disadvantages of psychologism."[5] Basic to the overcoming of psychologism is the distinction between psychic act and its logical content, the latter alone being the "in itself subsisting sense" ("in sich Bestand habende Sinn"). But can psychologism, which seeks to ground logic in psychology, be logically refuted? Perhaps not, Heidegger concedes in his dissertation, but that does not matter a great deal. He writes: "the actual...(also the non-actual) cannot as such be proved (*bewiesen*), but in any case can only be shown (*aufgewiesen*)."[6] While psychologism, according to Heidegger, as it is for Husserl, must be rejected, one needs nevertheless (i) to be clear about the real point of Husserl's critique of psychologism, and (ii) to decide where one should go after the error of psychologism has been discarded. For purposes of (ii), it is necessary (iii) to think about what is to be understood by 'Sinn', a concept which up until now has been used to define the domain of logic.

Part of Husserl's critique of psychologism in the *Prolegomena* relies upon a distinction between two modes of being, the real and the ideal. Thinking as a mental process is a *real* being; the logical content of each thinking has an ideal being. Psychologism confuses the two. The confusion does not lie in mistaking one given thing (the ideal content) for another given thing (the real mental process). It is rather based on the fact that the philosophers concerned were blind to, and prejudiced against, certain modes of being. So far Husserl's point was well taken. But Husserl's concept of 'ideal being' is far from being univocal. In fact, Husserl appears to have brought under this concept things that are very different from each other, such as universals, essences that are not universals, truths as well as the idea of truth. We shall look into some of these equivocations a little later. For the present, what is important in Husserl's critique, according to Heidegger, is not that ontological distinction which, however provisionally useful, could not be the final truth, but rather the implied critique of a naturalistic psychology. Hans Sluga has recently shown that when Frege rejected psychologism, he was in fact fighting against a more comprehensive philosophical naturalism of which psychologism was a consequence.[7] This reading is corroborated by Heidegger's understanding of Husserl's anti-psychologistic critique.

For Heidegger, it is a misunderstanding of Husserl's deeper intentions to read him as though he was improving upon Bolzano's platonism,[8] or even as though his critique was rooted in Lotze's *Geltungs-* and value-logic. These "platonistic" readings of *Prolegomena* have led to the standard complaint that in the second volume of *Logical Investigations* Husserl relapsed into psychologism. If we are to make room for the charitable interpretation that Husserl's *Logical Investigations*, even *Ideas*, constitute a progressive unfolding of the thoughts that were already anticipated in the early works, we have to say with Heidegger that Husserl rejected psychologism because it applied to logical theory a psychology which was not only poor *as a psychology* of the experience of thinking, but which was confused regarding its very project, which, in other words, did not understand its theme, i.e., the logical. The critique of psychologism therefore is a critique of psychology, and an implied plea for an intentional, descriptive, and eidetic psychology to replace the prevailing naturalistic psychology.[9] Such a reading of Husserl's intention makes it possible for Heidegger to go beyond the provisional distinction between the real and the ideal, and to ask how the logical contents or *Sinne* are related to the acts of thinking, and eventually to the thinking being that man is.

It is well known that Lotze's idea of *Geltung* or validity as the mode of being of propositions and truths influenced, in different measures, both Frege and Husserl. In his logic lectures of the twenties, Heidegger concerns himself at some length with Lotze. It is interesting to note that his assessment of Lotze underwent considerable change along the years. In 1912, Heidegger writes that Lotze's logic should be regarded as the basic book of modern logic.[10] In the Marburg lectures of 1925/26 we find him, in the course of a critical examination of Husserl's notion of 'ideal being', tracing Husserl's equivocations to the confusions that characterized Lotze's concept of

Geltung.[11] I will return to Lotze's concept of *Geltung* when we turn to the theory of truth. For the present it should suffice to note that amongst the entities whose mode of being is characterized by *Geltung*, Lotze includes: propositional contents or sentential meanings (= Frege's Thoughts), truths, the mode of being of a truth and the Essence of Truth. *Geltung* also means: *objective* validity (being true of objects) as well as universality with respect to all knowers. No wonder, then, that Heidegger severely criticizes those who find in this term "a magic band" capable of solving all problems.[12]

Heidegger was no more enthusiastic about Bolzano, the other major influence on Husserl. He cautions against regarding Husserl's *Logical Investigations* as nothing but attempts to improve upon Bolzano. It is, for him, more true to say that both Bolzano and Husserl were influenced by Leibniz. In any case, anti-psychologism does not lead Heidegger to the opposite camp of platonism. The goal is to be able to avoid platonism, without relapsing into psychologism.

C. Remarks on Mathematical Logic

For one who was so deeply concerned with traditional logic as Heidegger, the rise of mathematical logic could not but be a challenge. We know that Heidegger was enthusiastic about Frege's papers on the concept and object, and on sense and reference.[13] Of these he wrote: "G. Freges logisch-mathematische Forschungen sind meines Erachtens in ihrer wahren Bedeutung noch nicht gewürdigt, geschweige denn ausgeschöpft. Was er in seinen Arbeiten . . .niedergelegt hat, darf keine Philosophie der Mathematik übersehen; es ist aber auch im gleichen Masse wertvoll für eine allgemeine Theorie des Begriffs."[14] But the appreciation of Frege did not carry over into an appreciation of mathematical logic. In the same paper of 1912, he argues that logisitic -- as mathematical logic was alternately called -- does not liberate itself from mathematics and so is not able to penetrate into the proper problems of logic. Its chief limitations derive, in Heidegger's view, from an application of mathematical symbols and concepts (above all, of the concept of function) to logic -- as a result of which the deeper significance of the logical principles remains in the dark. As a calculus of propositions, it is unaware of the problems of the theory of judgment. Furthermore, the conditions of the possibility of mathematics, as well as of mathematical logic, lie in a domain which those two disciplines cannot reach.[15] In the Dissertation, a new objection is raised against mathematical logic: it is formal, and so is unable to deal with "the living problems of judgmental-meaning, its structure and its cognitive significance."[16] Similar complaints surface in later writings as well. In *Sein und Zeit*, logistic is said to "dissolve" judgment into a system of "Zuordnungen"; judgment becomes an object of "calculation," and so cannot be the theme for ontological interpretation.[17] Since judgment has always a relatedness to objects and a claim to be objectively valid, logistic cannot reach the essence of judgment.

Of what worth are these remarks? There is no doubt that Heidegger's acquaintance with the logic that Frege laid the foundation of, and that by the time Heidegger was writing his dissertation had found its epoch-making sytematization in Russell and Whitehead's *Principia*, was superficial and casual. Nevertheless, there may be some substance in his remarks.

That mathematical logic may well be so much of mathematics that it therefore becomes poorer as logic, is already implicit in Frege's criticism of Boole and Schröder. The point of that criticism is that Boole and Schröder used mathematical concepts ('sum', 'product', for example) and often mathematical signs to develop their logics, which is unjustified inasmuch as logic, being more fundamental, cannot and should not borrow its concepts from any other discipline.[18] Consequently, instead of reducing logic to mathematics, Frege reduced arithmetic to logic. He sought to make a fragment of mathematics logical, rather than make logic mathematical. It is true that Frege used at least two important notions in his logic which might be regarded as having been borrowed from mathematics. In fact, however, that is not so. Although the ideas of quantification and function are seemingly mathematical, they are not in reality. The mathematical notion Frege found confused and unhelpful. The logical notion that he introduced is that of any entity that is "unsaturated," i.e., has empty places within its structure. Thus a concept is a function inasmuch as its true form, on Frege's theory, is (for example) "______ is wise," and this is an incomplete entity. The same may be said of the quantifiers; they are, for Frege, properly logical notions, and not mathematical ones. Thus we must recognize that Heidegger's anxiety is genuine, but, as against the original Fregean logic, unfounded.

Heidegger's next complaint is that mathematical logic being a calculus of propositions, cannot raise the problems of judgment as discussed in traditional logic and metaphysics. What are these latter problems? As far as I can see, these problems are: (a) the nature of assertion/denial; (b) the nature of the copula and predication; and (c) the problem of truth. Limiting our view for the present only to Frege (and the logic of the *Principia Mathematica*, which is basically Fregean), we may say that Heidegger's critique is not justified if it means that Frege and the *Principia Mathematica* did not know of these problems. The only substance of the critique may be that the solutions offered by these new logicians were hardly satisfactory. Consistently with his critique of psychologism, Frege distinguished between assertion and the thought (or, in the *Begriffsschrift*, the judgable content, *beurteilbare Inhalt*) that is asserted. Thinking is grasping of the thought; judging is recognition of the truth value of the thought so grasped; and asserting is expressing that recognition. There is no doubt that the concept of assertion as a psychological (and linguistic) act and its relation (as well as that of grasping) to the thought (which on Frege's theory has an objective being) remains, in that theory, a "mystery" -- no less difficult to clarify than the role Frege assigned to 'assertion' in his logic, despite his anti-psychologism. These difficulties show that Frege's solution to the problem of assertion was not satisfactory, but there is also no doubt that he did concern himself with this aspect of the problem of judgment. As regards the problem of predication, which has been one of the central concerns of traditional logic and philosophy of logic, Frege's answer would run somewhat along the following lines: the problem of predication concerns the internal structure of the thought being asserted, and has nothing to do with judgment. Judging is recognizing the truth value of a total thought; the thought, or the judged content, contains a predicative structure, but even with regard to it one should note that what is the concept (or predicate) depends upon how one analyzes the thought and there is no one way of doing that. What about the copula? The copula as the connecting link between the subject and the predicate is no longer needed, for in 'Socrates is wise', the predicate is '______ is wise' and not 'wise'. This new way of analyzing a proposition better explains its unity than the copula does, for if the subject and the predicate were to be linked by a copula one may want to know what links the copula to both the terms, whereas on Frege's theory a thought consists of an "unsaturated" part (with a hole, as it were) and a "saturated" part (which just fits into that hole), each made for the other, and so not in need of a link.

What then is the point of Heidegger's remark that in mathematical logic, judgment is reduced to a system of *Zuordnungen* and not made a theme of ontological interpretation? If he means that modern logic looks upon a proposition as an unanalyzable primitive, then he is wrong. First-order propositional logic does so, but predicate logic precisely analyzes the proposition into its constituents. If he means a proposition is, for modern logic, a mere connection of concepts (or representations), then also he is wrong, for as Frege taught, a thought consists of a concept (or a function) and an object.

Further, the concept, for Frege, is not a subjective representation, but an objective entity. What then is the 'ontological interpretation'? It may mean either of four things: (i) interpretation of the fact that a judgment is about something, i.e., about a being; (ii) interpretation of the fact that a judgment is either true or false; (iii) interpretation of the mode of being of the judged content or proposition; and, finally, (iv) an answer to the question *how* something like a judgment is at all possible.

Of these four questions, Fregean logic has an account of (i) in terms of the object constituent of the referent of a thought; and an account of (ii) inasmuch as a sentence which expresses a thought also names a truth-value. Logicians such as Frege and Quine, to take two extreme examples, have ontologized about propositions or thoughts. The spectre of platonism has loomed large before them. It is not clear what is being asked by (iv). In any case, Heidegger's concern goes deeper than these answers. They are not radical enough both in their questioning and in their answers. With regard to (i), the Fregean answer does not succeed in locating the intentionality or object-relatedness of a judgment in the more general structure of intentionality, and gets by only with locating an object constituent. As far as (ii) is concerned, considering a sentence as a name of truth-value, in spite of the elegance it succeeds in bringing about in the semantics of first order propositional logic, does not question whether a sentence is after all a name,[19] and it demands an unquestioning acceptance of the very obscure ontology of the true and the false. It also does not, and indeed cannot, raise the deep question, Why is it that a judgment alone is capable of being either true or false? Taken together with a deep understanding of (iv), all these foregoing issues constitute what Heidegger calls 'philosophical logic'.

D. 'Philosophical Logic'

In his Marburg Lectures, Heidegger develops the notion of a philosophical logic as contrasted with the traditional "school" logic. The latter had its philosophical basis, no doubt, but now it is "der veräusserlichte entwurzelte und dabei verhartete Gehalt" of an original philosophical question. Philosophical logic has been developing through the centuries -- its high points are reached in Aristotle, Leibniz, Kant and Hegel. Amongst his contemporaries, Heidegger appears to have rated Lask most; he is the one who consciously strives toward a philosophical understanding of logic and sought to extend the domain of philosophical logic.[20] Husserl, in spite of the possibilities that phenomenology contained for a philosophical development of logic, did not succeed, in Heidegger's view, in conceiving logic philosophically: "he even intensified the tendency to develop logic into a separate science, as a formal discipline detached from philosophy." Nor did any other amongst the phenomenologists succeed. Pfänder's *Logik* -- widely regarded then as *the* phenomenological textbook on the subject -- is dismissed as "eine phänomenologisch gesäuberte traditionelle Logik."[21] Without pausing to evaluate these judgments on other philosophers (including those on Kant[22] and Hegel,[23] Bolzano[24] and Lotze), I will proceed to determine the tasks and the problems which Heidegger assigns to philosophical logic.

First of all, philosophical logic, as Heidegger conceives of it, is not a new discipline[25] but rather actualizes a *telos* which has characterized historical logic since its inception. The idea of philosophical logic, Heidegger claims, will first render the history of logic meaningful.[26] Philosophical logic, one may contend, can be brought about first by determining what philosophy is, and then by applying philosophy to logic. But where and how do we find the idea of philosophy to begin with? Heidegger prefers to follow another route. Let us begin with traditional logic (Aristotle or Leibniz, for example) and develop the central problems in it in such a manner that they will lead us into philosophy. We have no doubt a certain historical understanding of philosophy. With that much in our mind, we can question logic for its philosophical potentialities.

What are the problems that lead us from within traditional logic towards philosophy? These are:

1. Judgment, with which logic has ever been concerned, is characterized by intentionality; it is about an object, an entity. How to understand this intentional structure?[27]
2. What is the relation between the "being" of the copula and the "being" of ontology? How much ontological weight can we assign to the copula?[28]
3. What is predication and what role does it play in judgment?[29]
4. What is 'meaning', and what is its relevance for the possibility of judgment?[30]
5. What is the structure of judgment such that both the possibilities -- of truth as well as of falsity -- belong to it?[31]
6. How is truth related to judgment? Is it a property of judgment?[32]
7. Why is it that traditional logic has had two concepts of truth: propositional truth, and truth as self-evidence? How are these two concepts related? Are these legitimate concepts? What is their common presupposition, if there is any?[33]

8. There is a theoretical truth, as well as a practical truth. Which one of these is the primary sense of 'truth'?[34]
9. How is human thinking related to human existence?[35]
10. What is the metaphysical foundation of logic?[36]

To some of these questions we turn in the next parts of this essay.

2. THEORY OF JUDGMENT

A. Rejection of psychologistic theories of judgment.

In his *Dissertation*, Heidegger considers, in considerable detail, four theories of judgment -- those of Wundt, Maier, Lipps, and Brentano. Each of these theories is examined with regard to the general definition of judgment it gives; that definition is then tested by how it works in the cases of negative, impersonal, hypothetical, and existential judgments.

Of these four theories, Wundt's theory is concerned with the origin of judgment, Maier's with how a judgment consists of constituent act parts or *Teilakten*, and Lipps' with the completion of the process of judging. Brentano's comes closest to a purely logical theory, but still falls short of it.

(a) Wundt defines judgment as the analysis of a total representation (or thought) into its components. Judgment does not put together concepts, but rather analyses a thought into concepts. Of the latter concepts, the variable component is called the predicate, the relatively constant one is the subject.[37] Heidegger shows that Wundt's theory has no satisfactory account of impersonal judgments (such as "It rains"), existential judgments (the predicate "existence" is not given in the total representation that is analysed), hypothetical judgments (a ground-consequent relation cannot be extracted by analysis) and of negative judgments (Wundt does not in any case regard negation to be of special logical significance).[38]

(b) Maier rejects two common elements of the traditional theories of judgment: (i) the primacy accorded to the declarative sentence (*Aussagesatz*) as a grammatical entity (which, according to Maier, leads to the subject-predicate analysis that takes place under the misleading guidance of grammar), and (ii) the belief that 'true' and 'false' cannot be predicated of representations (*Vorstellungen*) themselves, but only of connections of representations. As against these, and in agreement with Brentano, Maier argues that judgment in its most basic form, is not a connection of representations. In "The sun shines,"

Judgements consist, according to Maier, of acts of presentation, which are then transformed into logical judgments by supervenient acts of objectification. An objectifying act is a positing of actuality, it is a sort of interpretive act. Besides these, there are two other component acts: an identification of the presently apprehended presentation with a reproduced one, and a *Wahrheitsbewusstsein*, which extends over all the three component acts.

Obviously such an account is a psychological, genetic account. The elemetary partial acts are generally, according to Maier, involuntary processes.[40] Against it, Heidegger asks: Is the primitive judgment of Maier the same as an elementary judgment in the sense of logic? Above all, Maier is concerned with the *act* of judging, not with the content of judging, the judgment as such. Logic has nothing to do with the processes, be they what they may, that might be "culminating" in the logical judgment. The logical judgment is not the completed final-state of the act; it is rather the objective content.

(c) Brentano, in common with Wundt and Maier, rejects the theory that judgment is a connection of representations. It would not do to say that the content of a judgment is complex, while the content of a representation is simple. The content of a judgment may be as simple as in "A is" (where one is not connecting "A" with "existence"); the content of a representation may be complex (as in the case of a question). This implies that, for Brentano, predication is not an essential component of judgment. What distinguishes a judgment from a mere representation is the presence of either recognition or rejection as a new manner of relatedness of consciousness to its object. Consequently, every judgment is existential, its object is being affirmed as existent or as nonexistent. Thus "Some one person is sick" translates, for Brentano, into "A sick man exists" and "No stone is living" into "A living stone does not exist".

Heidegger's criticisms of Brentano consist in showing in what sense Brentano's theory of judgment is psychologistic. Judgment is, for Brentano, a class of psychic phenomena. The *content* of judgment, that which is recognized or rejected, is of no interest to him. Thus while the distinction between the act and its content could have helped him to overcome psychologism, Brentano's interest remains with the psychic phenomena and he does not succeed in isolating anything specifically logical. It is true that his psychology being "eidetic," Brentano does not deny the universal validity of knowledge. But, as Heidegger insists, it is not a definition of psychologism to say that it denies the universal validity of knowledge.[41] The latter is at most a consequence of psychologism. What is important is that Brentano wants to ground logic in psychology. The act of recognition as such is not of interest to logic. The recognition must be justified. And the justification must lie in what is recognized. When one judges a > b (if a=5 and b=3), what is recognized is not the relation "greater than," but that the relation "holds good," its *Gelten*. This *Gelten*, "holding good," subsists independently of anyone's recognition.[42]

(d) Since Lipps' thinking underwent several major changes, he may be said to have held three different accounts of judgment. At first, he defined judgment as the consciousness of actuality (*Wirklichkeitsbewusstsein*), this consciousness being identified with a feeling of constraint (*Zwangsgefühl*). Next, he came to define judgment as consciousness of truth (*Wahrheitsbewusstsein*), where this consciousness is described as being constrained, in one's representation, by the represented objects (*im Vorstellen durch die vorgestellten Objekte genötigt zu sein*).[43] Finally, judgment came to be defined as consciousness of an object (*Gegenstandsbewusstsein*), where 'object' is distinguished from 'content' in that a content is sensed or represented, while an object is thought or meant and *demands* recognition. This demand or *Förderung* is a logical concept, as distinguished from the constraint or *Nötigung* (of the first two definitions) which is a psychological concept.

In Heidegger's view, Lipps' theory even in its final form remains psychological. Judgment is still an act, "my" response to the experience of *Förderung*. The 'feeling of necessity' even in the alleged logical sense should be kept out of logic.

The discussion concludes with certain remarks which point to further reflections. First of all, psychologism cannot perhaps be logically refuted. One can at most exhibit the peculiar nature of logical entities. If a logical entity is a *Sinn*, a *thought* (as distinguished from the act of thinking), then the essence of this entity is to be found not in a *Vorstellung*, but rather in the fact that it *alone* can *either* be true *or* false. It is to this last theme that much of the Marburg lectures of the late twenties are devoted.

Of the other conclusions Heidegger arrives at, some are more viable than others. I have already referred to his insistence that even if the logical entity has to be sharply distinguished from the mental process, the two must be set in some satisfactory relation. This, I think, is important. Both Husserl and Heidegger recognize this need, but pursue it along different paths.

Besides these two general conclusions which suggest further enquiry, Heidegger also proceeds to establish some specific conclusions. He, in a way, reestablishes the subject, predicate and copula analysis, as against its critiques by Wundt, Maier, Lipps, and Brentano. A judgment such as "a is equal to b" has to be construed as having 'a' and 'b' as subjects and "being equal" as pre-

some satisfactory relation. This, I think, is important. Both Husserl and Heidegger recognize this need, but pursue it along different paths.

Besides these two general conclusions which suggest further enquiry, Heidegger also proceeds to establish some specific conclusions. He, in a way, reestablishes the subject, predicate and copula analysis, as against its critiques by Wundt, Maier, Lipps, and Brentano. A judgment such as "a is equal to b" has to be construed as having 'a' and 'b' as subjects and "being equal" as predicate (as against the grammatical analysis which suggests 'a' as the subject and "is equal to b" as predicate). If the two-membered analysis holds good, then the copula is needed as a third component; it is just the relation between the two.[44] The copula, Heidegger admits, signifies not real existence, but mere validity (*Gelten*). It is in fact characterized as "something eminently logical," the most essential and proper element in a judgment.[45]

Logically more interesting is the next claim that the judgment relation has a certain irreversibility, a directionality, a *Richtungssinn*. Even in "a=b," equality holds good of 'a' and 'b', (and not that 'a' and 'b' of equality). By this, Heidegger rules out the possibility of different analyses of the same proposition.

As to negative judgments, he expresses dissatisfaction with the view that negative judgments are to be understood as judgments with negative predicates and refuses to regard a negative copula as an *Unsinn*.[46] In fact, negation, he adds, belongs originally to the copula,[47] and the two judgments, affirmative and negative, should be logically placed side by side.[48]

What about the impersonal judgment "It rains." The judgment, Heidegger insists, is not a naming judgment. It rather says, something *happens*, *takes place*, suddenly breaks in. The judgment, then, must be translated to "Raining is actual," "Of the raining, actuality holds good." He adds that this translation is unable to capture what we mean. The true meaning rather is something like this: "Of the raining, it holds good to take place now, the momentary existing."[49]

These are topics which have little influence on his subsequent concerns. So let me turn to his really continuing concern.

B. Judgment as the locus of truth and falsity

(a) Preliminary determination. If judgment is not a representation or a connection of representations, if its logical essence does not lie in its being a mental act, then we have to look for its essence elsewhere. It is generally agreed upon that judgments alone can be true or false. Perhaps it is here that we may be able to discern a clue to the nature of judgment, as also of logic. For logic alone deals with truth in general; the other sciences deal with truths.[50] And logic thinks about 'truth' only in connection with assertive sentences. Heidegger looks for some determination of the nature of such sentences, or of their meanings or propositions, which would account for *both* the possibility of being true *and* the possibility of being false.[51] Contrast Heidegger's problem with Frege's. Frege's problem was such that he could solve it simply by positing two objects which assertive sentences could *name*: i.e., the True and the False. This strategy works for the limited purpose of providing a semantic interpretation of propositional logic, but it leaves the main issue untouched. Are sentences in fact names at all? If they are not,[52] then what sort of structure must they (or their *senses*) have in order to be true *or* false?

The structure that Heidegger identifies is the opposition: putting-together (*Zusammensetzen*) and separating (*Auseinandernehmen*). The former is the condition of the possibility of truth and the latter, the condition of the possibility of falsity. But this is only an initial answer, and not quite correct. Not all affirmative sentences -- in which elements are put together -- are true, just as not all negative sentences -- in which elements are separated -- are false. The structure that is to be the condition of the possibility of both truth and falsity should consist both in putting-together and separation, in both at once.[53] What we need is a structure that is not merely a thinking together of the two surface structures of synthesis and separation, but which, being a unitary structure, precedes both.[54] We cannot think of this structure -- or even of putting-together and separation -- as a purely linguistic structure of the sentence. In the false judgment "The board is not black," the *words* are not more separated than in the true judgment "The board is black." Where then are we to look for this structure?

(b) 'Copula'. Perhaps it is in the "is" of the copula. We have seen that Heidegger does not go all the way with many of his contemporary logicians of different persuasions in rejecting the copula from theory of judgment. On the other hand, the precise sense of the "is" of the copula -- as distinguished from the "is" of assertion -- deeply interests him. In fact, as late as *Sein und Zeit*, Heidegger writes that the ontological significance of the copula has been lost to modern logic.[55] Logic since Aristotle has understood the copula as the sign for a combination of *ideas*, a combination that does not occur among things, but only in thinking. But at the same time, the "is" of the copula also signifies existence, essence (whatness), and truth or validity, in different contexts. (This ambiguity, we are assured,[56] is not a defect, but rather an expression of the intrinsically manifold structure of the being of an entity. This is a suggestion we need not try to understand for our present purpose.) What we need to focus upon is: what unitary structure of synthesis-cum-separation is to be discerned by reflecting upon the nature of the copula?

I do not think that Heidegger's logic lectures lead to any definitive answer to this question. But taking up hints from his writings, the following points may be singled out:

(i) In 'S is P', what is asserted is not bare identity, which would make it a tautology; nor is, for that matter, P different from S, which would have rendered the proposition necessarily false. There is thus a relation of identity-cum-difference.[57]

(ii) But what sort of things are S and P? They are not *Vorstellungen*, that was the point of the critique of psychologism. They are not words for obvious reasons. Are they Fregean senses or are they things? (Frege admitted both possibilities, but kept them apart. The sentence 'S is P' expresses a thought that is composed of the senses of 'S' and 'P'; but the sentence also has a reference that is composed of the referents of the component terms.) I think Heidegger's answer to this is much more complicated, and, if intelligible, profound.[58] *Logos*, in its totality, is a complex structure of words, meanings, the referent (what is thought) and what *is*. It is only when one separates them, that one seeks to tie them together by such relations as that of a sign to the signified. Verbal sound is not a sign for a meaning. Nor is the meaning a pointer to what is thought or to what is. There is an identity between these components,[59] an identity which yet shows the differences.

(iii) This last mentioned relational structure may be described as a structure of identity-cum-difference between thinking and being (where 'thinking' includes speaking, meaning and the meant, and 'being' includes being-as-referred, i.e., object and being as it is in itself). In judgment, thinking and being enter into a relationship. This makes it unacceptable to construe a judgment simply as a mental act directed towards a thought-content. Such a construal would set thought (as a timeless, abstract entity) apart from the world, and the act of thinking and expressing (as real, temporal events) from that thought. Thinking is not, as Frege would have it, grasping a thought, but thinking *about* a real being. I think one of the deep concerns Heidegger expresses in the Logic lectures is, how to articulate this *aboutness*, or intentionality of judgment.

With these three points (i) - (iii), we have already gotten some glimpse into the structure of judgment as involving both synthesis (identity, totality, involvement) and separation (difference, distinction). Traditional logic has not seen this interinvolvement of identity and difference, of thought and being, and on the basis of their absolute distinction, distinguishes between verbal and real propositions (Mill) or analytic and synthetic propositions (Kant). This latter sort of distinction has been questioned by many logicians in more recent times: by Quine, because no satisfactory criterion of synonymity is forthcoming, and by F.H. Bradley, earlier than Quine, because every judgment, in so far as it analyzes the totality of immediate experience, , is analytic, and in so far as it seeks to join together what analysis has torn asunder, synthetic. Heidegger's reason is different from both. The distinction between "the view of beings that makes itself manifest in common meaning and understanding, as it is already laid down in every language," and "the explicit apprehension and investigation of beings, whether in practice or in scientific enquiry" can hardly be maintained; one passes over into the other. In fact, the so-called verbal propositions, Heidegger insists, are but "abbreviations of real propositions."[60]

We still have to understand, how it is possible for a judgment to be about an entity. For Frege, it is so because the component name names an object (and the predicate refers to a concept under which that object falls). Heidegger's question is, how is that possible? Is he asking about the possibility of judgmental intentionality? To that, and some other related questions, we shall turn in the following part.

3. GROUNDING OF LOGIC

(a) Possible moves. There are various ways philosophers and logicians have sought to provide a "grounding" or foundation for logic. Starting with a logic, the most common move on the part of logicians, is to axiomatize it. This procedure will yield an axiomatic foundation. This is the most you can expect a logician qua logician to do. But in doing so, he is still doing logic, perfecting his logic, not "grounding" it in a sense in which philosophers have understood that task. Another move is to provide a logic with an ontological interpretation. In this case one starts with an uninterpreted system, and then assigns to symbols of appropriate types suitable entities belonging to appropriate types: such objects as singular entities and concepts, individual concepts, and propositions. One may thus admit various sorts of entities into one's ontology, or if one distrusts abstract entities, then he can use the semantics of possible worlds.

A more radical, and strictly philosophical grounding is called for when one asks about "the conditions of the possibility" of logic. How are logical entities such as judgments possible? How is it that formal logic is able to legislate the formal structure of any object whatsoever? Or, what are the conditions of the possibility of the objective validity and not merely formal validity of logic?

Faced with such questions, one may follow one of three possible paths. One may look for the transcendental foundation of logic in the structure of (human) consciousness; one may look for it in the structure of the world; or, finally, one may want to ground logic in man's intentional relationship with his world. The first is the path of Kant and Husserl, however different their conceptions of transcendental subjectivity, transcendental logic, and formal logic may be; the second is the path of platonistic metaphysics. Heidegger's path is the last one.

(b) Logic and Intentionality. In his habilitation work, Heidegger characterizes the nature of the logical thus: "The homogeneity of the domain of logic rests on intentionality, on the character of being-valid-of [*Hingeltungscharakter*]." Also: "Intentionality is the 'regional category' of the logical domain." He proceeds to explicate "intentionality" thus: There can be intentionality only in the case of what has meaning and significance, not in the case of what is just real.[61]

It would appear, then, that we can get at the roots of logic by following the guiding threads of this logical intentionality. This is what Husserl does in *Formal and Transcendental Logic*. But intentionality, for Heidegger, is not self-explanatory. It needs a "metaphysical" grounding, for which Heidegger argues throughout his writings. An intentional grounding of logic will show how the logical entities such as propositions, or the logical principles such as the principle of non-contradiction, are "constituted" in appropriate intentional acts. It will also show, as Husserl does in *Experience and Judgment*, how higher order intentional acts and their objects are built up on more primitive intentionalities and their objects. It should be noted that all this will be carried out within the scope of the transcendental epoche. The classical Kantian way is different, but also shares the same overall orientation. Formal logic has to be founded on transcendental logic, and transcendental logic lays bare the synthetic, world-constituting functions of the pure rational subject.

Once psychologism in philosophy of logic was rejected, two alternatives loomed large: the platonic hypostatization of the logical entities, and the Kantian-Husserlian thesis of "constitution" which, for one thing, respects the ideality of those entities, and, for another, sharply distinguishes the transcendental subjectivity from the psychological. Heidegger looked for a third alternative. But, in fact, he tries two different paths, and all his life sought to bring them together. One of these I will call the *metaphysical*, the other may be called the *practical*. The two are brought together in a *hermeneutic* thesis.

(c) Logic and Metaphysics. In the Logic lectures of 1928, called "The Metaphysical Foundations (*Anfangsgründe*) of Logic," Heidegger forcefully argues for the thesis that logic must be grounded in metaphysics.[62] Against such a thesis, there is a rather familiar objection which Heidegger considers at length. The objection is that since metaphysics involves thinking and since all thinking must conform to logic, indeed must presuppose logic, metaphysics must presuppose logic rather than the inverse thesis. Indeed, logic must precede all sciences.

According to Heidegger this argument has the advantage that it proceeds from quite general ideas of logic and metaphysics, without considering their specific problem-contents. There is also an ambiguity in the word 'presupposition'. It is true that all thinking -- prescientific, scientific, as well as metaphysical -- must make use of the formal rules of thinking. But *use* of the rules does not require a *science* of those rules, nor does it require a "founded" knowledge of those rules. The fact of their use, as much as the unavoidability of their use for thinking, needs to be accounted for. For such an account, one has to think about the conditions of the possibility of science, about the relation of science to scientific thinking, and of such thinking to human existence; logic itself is a science, historically developed and so determined by a tradition. It therefore cannot be a presupposition of thinking.

The barely formal argument to the effect that every thinking grounding must involve thinking, cannot be formally refuted -- Heidegger concedes.[63] But, he adds, it can be refuted only by showing how such an argument is possible and why, under certain presuppositions, it is indeed necessary. At this point Heidegger does not go on to show this. As far as I can see, his point would be something like this: pre-logical thinking which is in direct touch with being, thinking which, according to Heidegger's later writings, is either practical wisdom or poetic, does not follow the rules of logic and so no question arises about logic being its presupposition. It is only propositional thinking that follows the rules of (propositional) logic. A putative metaphysical grounding may remain within the limits of propositional thinking; it then does appear to presuppose logic (allowing for the sort of equivocation of "presupposing" which was hinted at earlier). Such a grounding then does not go to the roots of the matter. A metaphysical grounding which does go to the roots of the matter would think, but think in a different, more originary manner.

What is this more originary manner of thinking, and how could such thinking provide a grounding for logical thinking and for logic as well? To be able to understand Heidegger's answers to these questions, we need to do some more spade work to prepare the ground.

(d) Logic as Metaphysics of Truth.[64] Judgments alone can be either true or false. This is because in judgment, thinking and being enter into a peculiar relation of identity-cum-difference. Judgment is "about" a being, and of this being it asserts a true predicate. Let us look closer at this "being about" and also at the copula, the sign of predication.

(i) The "being about" or judgmental intentionality is possible, according to Heidegger, only because a being has already been disclosed prior to the judgment under consideration. A judgment does not first establish the relatedness to the entity-about-which. A judgment is first possible on the basis of an already available disclosure of the entity, and the disclosure of that entity takes place within the context of an already latent relatedness to, or *schon-sein-bei*, beings. A judgment is true if its content is in agreement with the already disclosed object-about-which. The metaphysical here is the disclosure of being as a being, a disclosure without which judgment cannot substantiate its truth claim and would not be, qua judgment, possible. Thus judgmental intentionality presupposes a prejudgmental manifestation of being. We need not have to understand this thesis in any weird or mystic sounding sense. The best way to understand Heidegger, at this point, is to take his thesis as exemplified in the familiar case of perceptual judgments. A perceptual judgment "This pen is blue" is possible inasmuch as the object-about-which, this pen, is already disclosed in perceptual experience, as lying there before me. It is important that we do not construe this perceptual disclosure itself as a judgment. What this disclosure is like, I will briefly touch upon later, but only in so far as that is necessary for my present exposition.

(ii) Predication likewise is founded upon display.[65] In predicating, what is disclosed is analyzed into one of its constituent moments, and this separated moment is exhibited as *belonging to* the entity disclosed. Predication determines an entity as being such and such, but the determination is founded on exhibition and separation. This shows why every judgment is both analytic and synthetic at once. The copula signifies the "togetherness," the "belonging-together," that "unifying gathering" which belongs to our very concept of being as the *world*.

(iii) If the foregoing makes sense, then it makes sense to say that although truth in the sense of adequacy or correspondence has its locus in judgment, truth in the sense of disclosedness of being is prior to judgment. If this latter sense of 'truth' be called ontological, then logic is grounded in ontology. Hence Heidegger's enigmatic statement: "Der Satz is nicht das, darin Wahrheit erst möglich wird, sondern umgekehrt, der Satz is erst in der Wahrheit möglich...Satz ist nicht der Ort der Wahrheit, sondern Wahrheit der Ort des Satzes.[66]

We thus find that when Heidegger claims to ground logic in metaphysics he should be understood in a sense that takes into account the above mentioned three points. He should not be construed as grounding logic either in the structure of the subject or in the structure of the world.

(d) Logic and Practical Wisdom. Logic, we have seen, deals with meanings. With the rejection of psychologism, one is tempted to look upon meanings as eternally subsistent entities. At no stage of his thinking was Heidegger satisfied with such a hypostatization of meanings. The habilitation work ends with the "metaphysical" suggestion that the opposition between real mental life and ideal meanings, between *Sein* and *Sollen*, be overcome in a more fundamental concept of living *Geist*.[67] The Logic lecture of 1925/26 suggests that although the primacy of theoretical truth is not accidental, it is possible to show that a more radical stance of questioning may lead to a revision of this naive point of departure of logic.[68] In fact, not formal logic but philosophical logic has to settle the question, which truth -- theoretical or practical -- is primary. Heidegger opts for the primacy of the practical.

To demonstrate this thesis of the primacy of the practical is to argue successfully that the meanings logic is concerned with, propositional meanings and their constituents, are *not* the meanings originally experienced along with that disclosure of being which is presupposed by judgment. The word-meaning, as *fixed* and stabilized for purposes of logical thinking, presupposes a pre-logical experience of being as meaningful. This latter sort of meaningfulness is tied to the way we live in our world and concern ourselves -- practically and affectively -- with things and situations. Things acquire their original significance (*Bedeutung*) from what we have got to do with them, from *Zutunhaben*. A pencil is meant for writing, a hammer for driving nails, and so on and so forth. Original practical judgments express such a significance of things: they do not ascribe properties to a thing. They are about my (actual or possible) relations to a thing.[69]

It may be objected that this sort of practical and affective significance belongs only to tools and artifacts: pens and pencils, houses and automobiles, hammers and clocks, but not to natural objects such as rocks and mountains, rivers and trees, and animals and other persons. I think Heidegger's point is that in so far as these and other natural objects inhabit my *Lebenswelt* and not the world of physics, they fall within the horizon of my interests, passions, and possible actions directed at them. They are not mere objects of cognition. The logic of judgment is founded upon the prelogical disclosure of things as having the sort of practical significance that they have within our *Lebenswelt*. To say this, however, is not to *show* how apophantic judgment arises out of the practical. It would be the task of hermeneutic logic to show that. Heidegger has not himself done hermeneutic logic; some others have, and we need to turn to them. But before doing that we need to be clear about how the practical wisdom which recognizes for each object and situation its practical significance could be characterized as being hermeneutic.

(e) Logic and Hermeneutics. It was said earlier that Heidegger tried, all his life, to bring together two different groundings of logic: the metaphysical and the practical, and that they were to be unified under the concept of hermeneutics. We now need to ascertain how this is done. The connecting link is provided by two theses: (i) that action is a mode of understanding the world and involves a certain self-understanding on the part of the agent; and (ii) that the originary disclosure of entities which must precede judgmental "being about" is not disclosure to a cognitive subject, to an objectivating consciousness, but rather to a projecting, caring, and acting being whose mode of being is to be in the world and to-be-already-with-entities. Being-in-the-world is to be interpreted as a certain comprehension or understanding of oneself and one's world. Thus both practice and disclosure of entities involve a certain pre-conceptual understanding of oneself and one's world. To articulate and explicate this understanding is hermeneutics. If logic is grounded in a disclosure of being, and if logical meanings refer back to pre-logical significance, one can as well say that logic is ultimately rooted in a certain understanding of the world as well as of oneself.

The same thesis may be supported in a slightly different manner. Judging is an intentional relation to a being. But every intentional relation carries within itself a *specific understanding* of the being or entity to which the intentionality relates. If judging presupposes a prior disclosure of that entity, it also requires a specific interpretation of it *as* such and such.

With this we are in a position to briefly consider Heidegger's thesis on logic as laid down in § 33 of *Sein und Zeit* bearing the title: "Die Aussage als abkunftiger Modus der Auslegung." In this paragraph, Heidegger first distinguishes between three meanings of "Aussage"; all three together constitute the full structure of *Aussage*. First of all, "Aussage" primarily means manifesting an entity as it is. In "The hammer is too heavy," the hammer itself, but not its representation, is manifested in the manner it is at hand. Secondly, *Aussage* also means predication. This sense is grounded in the first. Both the terms of predication, the subject and the predicate, belong to what has been manifested. Predication itself does not uncover anything but rather limits what has been uncovered to the subject, i.e., the hammer. Finally, *Aussage* also means "communication," to let the entity be seen together with an other. What is stated can be shared, can be stated again. Taking these three meanings together, an *Aussage* may be characterized as "communicating and determining, making manifest." But how then is it also a mode of interpretation? The making-manifest that takes place in and through an *Aussage*, is possible only on the basis of what is already disclosed to understanding. It is not a worldless, transcendental ego who performs an *Aussage*. It is rather a *Dasein* who is a being-in-the-world and as such always has a certain pre-understanding of the world, who makes a judgment. The existential fore-structures of understanding, which together constitute its anticipatory structure, form the horizon within which any judgment is possible. In this sense the judgment of logic is founded upon the hermeneutic of *Dasein*.

Heidegger has still to give an account of how the entity with which one is practically concerned (the hammer as a tool for driving a nail here and now) *becomes* an object about which one pronounces a theoretical judgment. Obviously, if Heidegger's thesis is correct, the *Zuhandene Womit des Zutunhabens* has to be transformed into the *"Worüber" der aufzeigenden Aussage*. What transpires in this transformation? Something whose mode of being is to-be-ready-at-hand becomes an object that is present-at-hand, merely *vorhanden*. The original "as," which was a hermeneutic "as" (recognizing a hammer as what is just right for my purpose) for practical wisdom, becomes a mere apophantic "as" (*judging* this object over there to be a hammer) which determines the object as possessing a certain property. The logic of *theoretical* judgments is committed to an ontology of objects present at hand.

In an important, but not much commented upon paragraph, Heidegger concedes that between these two extremes, there are many intermediate phases, represented by judgments about happenings in the surrounding world, accounts of situations, depictions of events, etc. These intermediate cases, though expressed in linguistic sentences, cannot be reduced to theoretical statements, but rather refer back to their origin in the pre-conceptual interpretation of the world.

What now has become of the concept of meaning or *Sinn* which was earlier used to define the domain of logic? This concept of *Sinn* is to be traced back to its origin in another, more originary concept of *Sinn* which Heidegger formulates with some precision in § 65 of *Sein und Zeit*: Danach ist Sinn das, worin sich die Verstehbarkeit von etwas hält, ohne dass es selbst ausdrücklich und thematisch in den Blick kommt. Sinn bedeutet das Woraufhin des primären Entwurfs, aus dem her etwas als das, was es ist in seiner Möglichkeit begriffen werden kann." *Sinn* is that towards which the originary project of being-in-the-world is directed. To understand the *Sinn* of a thing (not of a word, in this case) is to grasp, unthematically, the possibility that the thing presents in the context of the prevailing project.

(f) Hermeneutic Logic. It is one thing to claim that formal logic is rooted in a hermeneutic experience of being-in-the-world. It is quite another thing to work out in detail the idea of a hermeneutic logic. Without such a logic, the Heideggerean thesis would remain empty of content, for not only logic but all theoretical cognition, on that thesis, would have the same "origin." With such a logic, the thesis receives specific content, but loses some of its ontological grandeur, for now formal logic will be traced back to another kind of logic, but we would still be within the field of logic, which thereby would receive an extension beyond the formal-theoretical.

Even if Heidegger does not give us sketches of such a logic, luckily we have excellent attempts in that direction. This is not the place to review those attempts, but it surely is appropriate that we briefly recall the more noteworthy amongst them. First of all, Husserl himself, in *Experience and Judgment*, extended the domain of logic to pre-predicative experience, and showed how truth-functional operators such as negation, disjunction and implication have their origin in pre-predicative experience. Husserl's thesis may be regarded as still being cognitive in nature, the pre-predicative experience is construed not as active or affective dealing with entities, but rather as modes of receptivity and various modes of responses to what is received. In this sense, Husserl's pre-predicative logic does not come under the rubric "hermeneutic logic."

The most striking development of hermeneutic logic, developed in close contact with both Husserl and Heidegger, is to be found in the works of Hans Lipps.[70] If formal logic deals with logical entities which claim to be self-subsistent essences, and appear to have no connection with the living situations of everyday life, what Lipps does is to comprehend precisely the entities and structures of logic as arising out of human life, i.e., to bring out how they originally have the function of accomplishing quite specific roles in quite specific linguistic situations in everyday life. Thus judgment (*Urteil*) in its origin is not a statement, but an action by which a yet-to-be-decided question is finally decided, as in legal judgment. The concepts of traditional logic, according to Lipps, are quite different from the concepts of originary, practical thinking. To comprehend things, in practical life, is to come to terms with things, to know what to do with them, as in overcoming an opposition. Concepts in this sense are not definable, they can only be illustrated by examples. The same sort of distinction is made in the case of inference. In practical life one infers, not from premises, but from circumstances, situations, facts. *Proof* becomes necessary in a situation of conversation, when something has to be demonstrated for the other. An interesting development of the idea of pre-logical conception is Lipps' distinction between "practical" and "intuitive" (*sichtenden*) conceptions. Neither needs language, but both may function in a linguistic medium. The practical conception operates in knowing *how*; the intuitive conception operates in one's mastery over a wide range of diverse material without yet subsuming it under a common logical concept.

Meanings of words are, for theoretical logic, precise and fixed entities. In practical life, meanings cannot be fixed with precision. (Lipps elaborates on the Wittgensteinian example: the word "game.") This imprecision is not a deficiency; it is rather a strength. The words derive their meanings not autonomously, but in connection with situations in which they are uttered. This leads Lipps to consider various kinds of words and the great variety of situations that call forth appropriate utterances.

Josef König studied with Husserl, but subsequently attended Heidegger's Marburg lectures, and sought to appropriate their methodologies into a basically Dilthey-oriented position. I would here mention only a few of his important distinctions: (i) In his *Sein und Denken*[71] König distinguishes between the merely present (*vorhanden*) thing and the thing as so-working (*so-Wirkende*). The former is not an original subject of predication, but is rather a transformation of a judgment of the form "X is present." The true subject of a statement about something present is not this something present, but rather the X of sentences of the sort "X is present." But the latter, i.e., the so-working, or an entity that is not the merely present, is the original entity. The subject of so-working is *nothing but* a so-working being (a pleasing smile is a smile that so works on us; a sublime mountain is one which so works on us). Its being (*Sein*) is to be so-working.

(ii) König also distinguishes between a practical 'this' and a theoretical '*this*'.[72] The theoretical *this* is a *this* of such and such kind: for example, 'under this circumstance' = 'under *such* circumstance'; this man = a man such as this. As contrasted with this, a practical *this* is a pure this. For example, What is *this* that lies there on the table? A practical this is the merely existing reality. The practical this belongs to someone's world; it is hardly compatible with the thought of a closed system or with a worldly-totality as *Vorhanden*.

(iii) Another of König's related distinctions is that between practical cause and theoretical cause.[73] The former answers a practical "why" question and the latter a theoretical question. A practical "why" question is: "Why does this ball start moving?" A theoretical "why" question is "Why do balls that receive an impact start moving?" The former is answered by giving another event as the efficient cause. The latter requires a ground in a general theoretical implication.

(iv) All these lead him finally to a distinction that is of direct significance for logic: that between practical sentences and theoretical sentences.[74] A theoretical sentence (or proposition) can be rightly seen as built out of a sentential (or propositional) function 'x is F' either by replacing 'x' with a constant 'A', or by quantifying over x (Some x is F; All x is F). A practical sentence, according to König, cannot be so construed without doing violence to its meaning and its role. The subject of a practical sentence is a practical "this" or "that." The sentence, "That is my friend Karl" cannot be regarded as having been built out of a sentential function "x is my friend Karl."

König's valuable, carefully developed, but incomplete researches shall constitute a necessary part of any satisfactory hermeneutic logic.

Lastly, I should mention the more well known and more recent attempt of Paul Lorenzen.[75] Lorenzen develops a systematic constructive procedure for building up formal logical concepts and operations from simple situations (such as one in which one person gives an order which the other obeys or does not obey; or one in which two are engaged in a game; or dialogical situations in which there is a proponent and an opponent). Lorenzen, interestingly enough, sees his task as having been made possible only after Dilthey and Frege.[76]

One may want to say that these attempts fulfill the intention implicit in Heidegger's thinking about formal logic, in a more constructive and fruitful manner.

4. CRITICAL REMARKS

But what to say about Heidegger's own foundational thoughts? To recapitulate what has already been pointed out, these thoughts are mainly five:

Firstly, formal logic, *historically*, was possible within a metaphysical system (the Platonic), and can be made possible only within a metaphysics.

Secondly, formal logic is committed to an ontology of objects whose mode of being is to be present at hand (*Vorhandensein*).

Thirdly, (in spite of the above) philosophical reflection on the copula yields an insight into the identity-cum-difference, and the togetherness of differentiated elements that belongs to the *meaning* of Being.

Fourthly, judgmental being-about presupposes a prior pre-judgmental disclosure of an entity, which disclosure takes place within the context of Dasein's already-being-with the others.

Fifthly, judgmental *Sinn*, as also logical-theoretical meaning of words, refers back to a practical understanding of the significance of things in relation to human projects, i.e., in the context of the totality of life situations.

The final evaluation of formal logic would be somewhat as follows: formal logic has its own range of validity, no doubt, but philosophy should replace its naivete by reflecting on its sense and its "origin." This will require a philosophical logic which is double-pronged: at once ontological and hermeneutic. Modern mathematical logic is degenerate formal logic, for whatever hermeneutic and ontological glimpses the traditional formal logic permitted is, or at least appears to have been, totally lost to mathematical logic, whose main blunder consists in confusing between a science of quantity and a science of intentionality and which is, historically speaking, possible only in an epoch for which the meaning of Being is understood through technology.[77]

With regard to these thoughts, I would like to submit the following critical and, certainly, tentative reflections.

1. The historical judgment appears to me to be sound, namely, that formal logic arose within the Platonic metaphysics. One needed, to begin with, a doctrine of objective ideas and propositions. But the history of logic shows that logic has tried to free itself from that Platonic origin. Propositions have been replaced by sentences (even if they are 'eternal sentences'), concepts by words (even if they are type words, not tokens), and so on and so forth. To what extent, then, must we say that formal logic unavoidably presupposes a metaphysics (i.e., a theory of Being) and an ontology (a position as to what sorts of entities to admit)? My own view is that although formal logicians have

sought to court a nominalistic ontology, that just has not worked. (See how sentences have become eternal sentences.) The logical relations and structures need abstract entities to hold good of, so some sort of Platonism is 'the original sin' of formal logic. But these Platonic entities are of the genre of meanings, Fregean *Sinne* or Husserlian *noemata*. A certain theory of meaning, and its attendant ontology may well be regarded as the minimum commitment of formal logic. No other ontology of *Vorhandensein* is presupposed. Events and happenings, situations and circumstances, tools and gadgets, can all be referents of "objects-about-which" of *propositions* that are subjected to logical operations.

2. It is not clear how much ontological burden can be carried by the copula. Heidegger's multifarious attempts to extract out of it insights into the meaning of 'Being' have been far from successful. By saying that 'Being' involves identity-cum-difference or the togetherness of distincts, is not to say much that could not be divined by simple metaphysical speculation independently of the guidance of the copula.

3. The thesis of the pre-logical, pre-predicative disclosure is important, and its validity should be recognized. I should add that this thesis derives its strength from the case of perceptual judgments such as "This pencil is blue." But not all judgments are perceptual, and not all disclosure is prior to judgment. In a judgment about electrons, one does not have a pre-theoretical disclosure of the object-about-which: in verifying such a judgment, the disclosure comes *afterwards* as the "fulfillment" of the meaning intention of an originally empty judgment. The thesis of *prior* disclosure, then, may be saved by liberalizing the sense of 'disclosure' and at the same time by relativizing it to the context of a judging.

4. With regard to perceptual judgments about persons and material objects, it is true that originary disclosure is not a theoretical-cognitive mode of givenness, but rather practical and affective.[78] This alone justifies Heidegger's basing apophansis on hermeneutics. However, even if one does work out a hermeneutic logic in the manner of Lipps, König, and Lorenzen, one still needs to show how apophantic logic develops out of hermeneutic logic. Lorenzen's is the best attempt to show this, but it works for elementary truth functions, and even there a certain *discontinuity* between the primitive hermeneutic situation and the formal-logical is either slurred over or eliminated by choosing the former at a level that is not originary-practical, but rather primitively theoretical.

5. Heidegger is right, to my mind, in looking upon Husserl's anti-psychologism critique as a provisional, though indispensable step. In fact, Husserl himself treated it likewise. The gap between real mental life and ideal meanings has to be bridged. Transcendental philosophy and hermeneutics are two ways of doing this. Their relative strength has to be measured, among other things, by the extent to which each is capable of accounting for the ideality of logical meanings. For hermeneutics, the question is: How do the practical-hermeneutic meanings of things get 'transformed' into the theoretical-logical meanings of words and sentences?

NOTES

[1] *Die Lehre vom Urteil im Psychologismus. Ein kritisch-positiver Beitrag zur Logik.* Dissertation, Freiburg in Br. , 1913. Reprinted in Martin Heidegger, *Gesamtausgabe*, Bd. 1, *Frühe Schriften* (Frankfurt am Main: Klostermann, 1978). Further citations to *Gesamtausgabe* are abbreviated as GA.

The reader is referred to the following secondary literature on this topic: Thomas A. Fay, *Heidegger: The Critique of Logic* (The Hague: Martinus Nijhoff, 1977). Reviewed by the present author in *The Southwestern Journal of Philosophy,* XI (1980): 174-79; Walter Bröcker, "Heidegger and Symbolic Logic," in M. Murray, ed. , *Heidegger & Modern Philosophy* (New Haven: Yale University Press, 1978), 3-22.

[2] *Die Kategorien- und Bedeutungslehere des Duns Scotus* (Tübingen: J. C. B. Mohr, 1916). Reprinted in *Gesamtausgabe*, Bd. 1.

[3] *Logik: Die Frage Nach der Wahrheit.* Vorlesungen 1925-26, herausgegeben von Walter Biemel. *Gesamtausgabe*, Bd. 21 (Frankfurt am Main: Klostermann, 1976); *Metaphysische Anfangsgründe der Logik*, Vorlesungen, 1928, herausgegeben von Klaus Held, *Gesamtausgabe*, Bd. 26 (Frankfurt am Main: Klostermann, 1978).

[4] GA, 1:186.

[5] GA, 1:20.

[6] GA, 1:165.

[7] Hans Sluga, *Gottlob Frege* (London: Routledge and Kegan Paul, 1980).

[8] GA, 1:87, fn 9.

[9] GA, 1:98.

[10] GA, 1:23 fn.

[11] GA, 21:62.

[12] GA, 21:79.

[13] G. Frege, "Begriff und Gegenstand," *Vierteljahresschrift für wissenschaftliche Philosophie*, 16, (1892); "Sinn und Bedeutung," *Zeitschrift für Philosophie und philosophische Kritik*, 100 (1892).

[14] GA, 1:20.

[15] GA, 1:42-43.

[16] GA, 1:174 fn.

[17] Heidegger, *Sein und Zeit*, Seventh edition (Tübingen: Max Niemeyer, 1953), 159.

[18] Cf. Frege: "Anyone demanding the closest possible agreement between the relations of the signs and the relations of the things themselves will always feel it to be back to front when logic, whose concern is also the foundation of arithmetic, borrows its signs from arithmetic. To such a person it will seem more appropriate to develop for logic its own signs, derived from the nature of logic itself. " *Posthumous Writings*, ed. by H. Hermes, F. Kambartel and F. Kaulbach (Chicago: University of Chicago Press, 1979), 12.

[19] cp. M. Dummett, *The Interpretation of Frege's Philosophy* (London: Duckworth, 1981), 371, 409.

[20] M. Heidegger, *The Basic Problems of Phenomenology*. Marburg Lectures of 1927, Ed. and trans. by Albert Hofstadter (Bloomington: Indiana University Press, 1982), 178. Henceforth to be cited as BP.

[21] GA, 21:28.

[22] Kant, according to Heidegger, gave logic a central philosophical function but did not try to rescue academic logic from its "philosophically alienated superficiality and vacuity. " (BP, p. 177).

[23] Hegel, Heidegger holds, conceived of logic as philosophy, but did not attempt a radical reformulation of the problem of logic as such. (BP, pp. 177-8).

[24] Bolzano, in Heidegger's view, was overrated by Husserl. (GA, 21, pp. 86-7).

[25] GA, 26:6.

[26] Ibid, 7.

[27] GA, 26:158f.

[28] BP, 177, 211f; GA, 26:26f.

[29] BP, 208ff.

[30] GA, 26:151f; *Sein und Zeit*, 148f, 216f.

[31] GA, 21: 134-50.

[32] GA, 26: 125-26.

[33] GA, 21: 110-129.

[34] GA, 21: 11-12.

[35] GA, 26:24.

[36] GA, 26: 170, 128ff.

[37] Contrast Frege who regarded the predicate part or the function as "the stable component" and the sign for the object, i. e. , the argument as replaceable by others. Cf. *Begriffsschrift*, 9.

[38] Again compare Frege who regarded the distinction between affirmative and negative judgments as "eine für Logik wenigstens ganz unnötige Unterscheidung, deren Grund ausserhalb der Logik zu suchen ist. " ("Verneinung," reprinted in Frege, *Logische Untersuchungen*, G. Patzig, ed. (Göttingen: Vandenhöck & Ruprecht, 1966), 61.

[39] Cf. Russell's thesis that "this" is a proper name, together with Husserl's thesis that the naming act may be true or false.

[40] Heidegger does not consider, in his critique of Maier's theory, a possibly transcendental-psychological interpretation of the theory in the sense of Kant's doctrine of three-fold synthesis.

[41] GA, 1:122.

[42] GA, 1:123f. Compare Frege's view that judgment is the recognition of the truth value of a thought.

[43] Quoted by Heidegger in GA, 1:135.

[44] If only Heidegger had construed the predicate not as "being equal," but as "_______ is equal to _______," then he would have realised Frege's point that the names of the so-called subject terms just fill in these blanks, and so no third connecting link is necessary.

[45] GA, 1:178-79.

[46] GA, 1:183.

[47] GA, 1:184.

[48] GA, 1:185.

[49] GA, 1:186.

[50] GA, 21:7. Compare Frege: "The word 'true' can be used to indicate such a goal for logic. . . of course all the sciences have truth as their goal, but logic is concerned with the predicate 'true' in a quite special way. " *Posthumous Writings*, 128.

[51] GA, 21:135f.

[52] Dummett rejects this part of Frege's semantics.

[53] GA,21:136f.

[54] GA, 21:140-41.

[55] *Sein und Zeit*, 159-60. Also see 349.

[56] BP, 204-205.

[57] Many Hegelian logicians, such as F. H. Bradley, have used this so-called paradox of predication to imply that judgmental thinking cannot know reality. One may, contrariwise, regard the puzzle as signifying that structure which makes both truth and falsity possible.

[58] BP, 207.

[59] Husserl's sixth logical Investigation has texts which suggest such a view, cf. §§ 6-7.

[60] BP, 197.

[61] GA, 1:283.

[62] GA, 26:128-32.

[63] Ibid. , 131.

[64] GA, 26:132.

[65] BP, 209f.

[66] GA, 21:135.

[67] GA, 1:405.

[68] GA, 21:11.

[69] GA, 21:150-59.

[70] Hans Lipps, *Untersuchungen zur Phänomenologie der Erkenntnis*. Erster Teil, *Das Ding und seine Eigenschaften* (Bonn, 1927). Zweiter Teil, *Aussage und Urteil* (Bonn, 1928). But more specifically, see his *Untersuchungen zu einer hermeneutischen Logik*, Philosophische Abhandlungen, Bd. VII (Frankfurt am Main, 1933).

[71] Halle: Max Niemeyer.

[72] Josef König, "Über einen neuen ontologischen Beweis des Satzes von der Notwendigkeit alles Geschehens," *Archiv fur Philosophie*, 2 (1948): 5-43. Reprinted in Josef König, *Vortrage und Aufsätze*, ed. G. Patzig (Freiburg/München: Verlag Alber, 1978).

[73] Josef König, "Bemerkungen über den Begriff der Ursache," originally in *Das Problem der Gesetzlichkeit*, Bd. I (Hamburg: F. Meiner, 1949). Reprinted in *Vorträge und Aufsätze.*

[74] König's Göttingen Lectures (1953-54) under the title "Theoretische und praktische Sätze" are still unpublished. They are being edited by G. Patzig for publication.

[75] Cf. Paul Lorenzen, *Konstruktive Wissenschaftstheorie* (Frankfurt am Main: Suhrkamp, 1974) and *Methodisches Denken* (Frankfurt am Main: Suhrkamp, 1974).

[76] *Konstruktive Wissenschaftstheorie*, 21. He also writes: "Erst im Anschluss an Dilthey und Husserl haben Misch einerseits und Heidegger andererseits deutlich gemacht, was das heisst, dass Denken vom Leben, von der praktischen Lebensituation des Menschen, auszugehen hat. " *Methodisches Denken*, 26.

[77] For my present limited purpose, I desist from either expounding or commenting upon this last claim.

[78] I have argued for this in my *Phenomenology and Ontology* (The Hague: M. Nijhoff, 1970).

7. JOSEF KÖNIG'S DISTINCTION BETWEEN THEORETICAL AND PRACTICAL SENTENCES*

Josef König's thinking is dominated by the idea of "radical" difference and by a particular example of that radical difference. I will begin with a few preliminary words about both.

Two things A and B may be either simply different, i.e. different Φ s; or they may be radically different, i.e. not only different Φ s, but also *as* Φ s different. Thus Plato and Aristotle are different men, they are different instances of the generic concept "man". They are not however as men different. They are men in the same sense. However, to recall a famous Aristotelian thesis, if physical substances, numbers, universals and God are all beings, they are not merely different beings, they are as beings different. The very sense of their beingness is different. In still other words, they are not to be regarded as different instances of the generic concept "Being", the concept "Being" which applies to them all is rather an analogical concept. It is very important for philosophy in general, but more so for König's thinking, that this formal notion of radical difference be kept in mind, so that when two things are radically different we do not treat them as simply different.

König's lectures and writings abound in examples of such radical difference. Let me recall some. Husserl's distinction between "generalisation" and "formalisation"[1] is one such: they are not species of a genus. A formal generality such as "something" is a radically different generality than a material essence such as "redness". In his early work on the concept of Intuition[2] König insists that sense-perception and intellectual intuition are, as intuitions, different. Brentano's "modifying predicates" and "determining predicates" are likewise radically different sorts of predicates. In his Göttingen inaugural lecture on "*aesthetic effect*", König advanced the thesis that description of an aesthetic effect (as when Goethe says[3] that an art work brings about a sort of stability and security) and description of a non-aesthetic effect ("Hearing that news makes me feel good") are, as descriptions, different.[4] In his *logic lectures,*[5] he says that if there is an expression α , of which it holds good that

* Originally appeared under the title, "The Central Distinction in Josef König's Philosophy" in *Dilthey-Jahrbuch*, Bd. 7, 1990-1991, pp. 230-249

different αs are *eo ipso* radically different, such an α = "world". A world is *eo ipso,* for those living in it, *the* world. From inside, "world" is not a genus of different worlds.

There is, however, one case of radical distinction which runs through König's work, and which may be regarded as his most important contribution to philosophy. This is the distinction between theoretical and practical sentences. The two-semester long lecture bearing that title is still unpublished, and a longer manuscript in a proposed *Festschrift für Georg Misch* never appeared.[6] The following account of that distinction shall make use of my own notes from the *logic lectures* and also of the *Misch Festschrift* paper. My plan is to give a preliminary account of the distinction (Part I), then to return to works published before the distinction explicitly appeared in order to trace its links to them (Part II), and finally to return to this distinction in order to develop its wider implications and to raise some critical questions with regard to it.

I

In the *logic lectures,* König begins by considering the traditional doctrine of division of sentences according to their quantity into universal and particular on the one side and singular on the other. Lukasiewicz regards it as a flaw in Aristotle's theory to have neglected singular sentences.[7] For König, on the other hand, Aristotle's distinction is of the utmost importance. But Aristotle regarded both — i.e. universal and particular sentences on the one hand and singular sentences on the other — as species of the genus *apophantic logos* ("logos" being defined as an expression which means something and whose parts also have the capacity to mean something). Following Aristotle, traditional logic regards them as belonging to the same genus, and does not distinguish them as sentences. They are taken to differ only with regard to their quantity.

Apophantic logos however is distinguished from non-apophantic logos by being an expression which means something insofar as something is or is not. This it does by affirming or denying something of something, by virtue of which it is essentially either true or false. Non-apophantic sentences are those of which none of these three features — features which are sufficiently closely connected to form one inter-connected structure — holds good. Examples of non-apophantic sentences are: "Please pass on the sugar" (a request) and

"Close the door" (an order). Of these, obviously, truth and falsity do not hold good.

What König proposes — as a first approximation — is to put together the singular sentences such as "Socrates is mortal" and "That is my friend Karl" and Aristotle's non-apophantic sentences such as those expressing requests and orders under a common heading "practical sentences". Under "theoretical sentences", then, fall all Aristotle's apophantic sentences excepting the singular ones. We still have to understand why he insists that these two groups of sentences are radically different i.e. different with regard to their very character as sentences. Before we return to this question, it should be noted that what he calls "practical sentences" includes two heterogeneous groups:

singular and demonstrative sentences on the one hand and sentences expressing requests, commands etc. on the other. What is common to them besides the fact that they both are not to be classified as either universal ("All men are mortal") or particular ("Some Indians are honest") sentences? Using König's own locution, can we ask: are these two merely different practical sentences, or are they different *as* practical sentences? Postponing consideration of such questions, let us pursue our immediate interest to explain the larger distinction which König presses with such vigor and philosophical rigor.

Now, following Aristotle, König regards both universal and particular sentences as showing that something is true of a universal thing (both "All men are mortal" and "Some men are honest" show that something is either always or sometimes true of the universal thing *Man*), while singular sentences show that something is true of a singular thing. In this way of drawing the distinction, the important distinctions are between "universal thing" and "singular thing", and also between the two different senses of "showing". A universal thing, according to Aristotle, can be predicated of many, whereas a singular thing cannot be predicated of anything. (In "That is Karl", what is being predicated is "called Karl".) For our present purpose, we need not raise the question, if Aristotle's universal thing *Man* is a thing or a word. What is important for our purpose is to note the difference between "man" and "Karl". But this is only part of the distinction between theoretical and practical sentences. The other part lies in the different ways they show something to be the case. In the *logic lectures*, König asks: who is the subject of this "showing" in each case? In the case of "Karl is blond", shall we say that the sentence itself shows whatever is shown in it? According to König, a sentence itself shows, when anyone who understands it can, on the basis of such understanding, know the thing that is predicated and the thing of which it is predicated. In the case of "Karl is blond", this is not the case. In order to know of whom being-blond is predicated, it is not enough that one understands the sentence, one also needs to know who Karl is. The sentence "shows" what is true of something, but does not show of what that something is true. As contrasted with singular sentences, a theoretical sentence just shows what it says, it shows what is predicated, and of whom. Anyone who understands the sentences "All men are mortal" and "Some men are honest" also knows both what is being predicated in each case *and of whom.*

Another way of putting the same point across is this: one who understands a theoretical sentence is also "immediately" (i.e. without further questioning) capable of asking, is it true or false? This is not to say that he can decide this question satisfactorily, "immediately" or as a result of further investigation. What is relevant for our present purpose is that the question, whether the sentence is true or false, can be meaningfully asked simply on the basis of one's "immediate" understanding of the sentence. One who understands it knows what is to be there in order for it to be true. In König's words, theoretical sentences, when they are true or false, are by themselves (*selber*) true or false. It is quite otherwise with a singular sentence such as "Karl is blond". Understanding it does not "immediately" tell us which Karl is being meant. To ask, whether the sentence is true or false, is meaningful only when one knows, after further questioning, who this Karl is. Again, in König's words, singular sentences, when they are true or false, are not so by themselves (*selber*). The singular sentence "Karl is blond" is true or false only on the condition that the person who utters it has a definite Karl in his mind. A theoretical sentence, on the other hand, is true or false without any further such condition. This is the same as saying, if p is a theoretical sentence, and if q be the sentence "p is a true sentence", then p≡q. This equivalence does not hold good of a singular sentence. Still another way of expressing the same distinction is this: a theoretical sentence itself states something. What we state by uttering it, is what the sentence itself states. What we mean by it is what the sentence itself means. A practical sentence, in König's view, does not itself state anything. It states or means what the utterer states or means by uttering it. Like a vector quantity, it is directed towards someone.

When I tell you "All men are mortal", what I communicate to you is a proposition, a Fregean thought. When I tell you "Karl is blond", what I communicate to you is not a proposition but the fact that Karl is blond, or rather Karl's being blond. I am talking to you about Karl himself, not about "Karl". The word "man" is "immediately" meaningful, the word "Karl" can mean something only when someone means some definite thing by it. The meaning of a practical sentence is thus situation-bound, it does not express a Fregean sense. We shall return to that important idea later in this paper. For the present, now that we have a preliminary exposition of König's distinction between theoretical and practical sentences, let us look at some of his other theses — centering around similar distinctions — in three of his other works: first, his *Habilitationsschrift* entitled *Sein und Denken*,[8] next in his major article on the *concept of cause*,[9] and finally in his Göttingen inaugural lecture on *the nature of aesthetic effect*.[10]

II

So, first, let us turn to *Sein und Denken*, a most difficult book to read, partly because of a complex system of technical vocabulary it uses, but also because of a very difficult mode of thinking it embodies. Starting from Brentano's distinction between determining and modifying predicates (which, following König's locution, are as predicates different), the discussion focuses upon a group of modifying predicates which are said to express the *how* of a certain mode of working (*Wirken*). To say of a thing that it is beautiful is to describe how the thing is working on the speaker. A beautiful thing is not a thing which is already there, but only now works as beautiful. It is rather nothing but a beautiful-working thing. In this sense, being red, for example, is extrinsic to the being of a thing that is red, but the being beautiful is intrinsic to the being of a thing which works as beautiful on someone. A beautiful thing i.e. a thing working upon us as beautiful, is what it is by producing on us an impression *of* beauty. This impression *of* beauty (or, *of* vitality or *of* sublimity) is a determinate impression, but beauty or vitality or sublimity as the case may be, is not an attribute of the impression. When one speaks of a clear or a fleeting impression, one is describing the attributes of the impression. But an impression *of* clarity does not have clarity as its attribute. "Sublime" is never an attribute of an impression. The "of sublimity" or "of beauty" gives the *Wovon* of the impression — its immanent object. This relation is different not only from that between an impression of red and the red of which it is an impression. It is also different from a fingerprint's relation to the finger of which it is the print. A finger is already known prior to its print. But beauty or sublimity (or, to take another example, spatial depth) is originally presented only through an appropriate impression — impression of beauty or of sublimity (or of spatial depth). The Humean impression is like a fingerprint. An impression of beauty manifests beauty, beauty resides in it as its immanent object (not as red resides on the surface of a red paper). Perhaps, one should best speak of "impression-of-beauty".

Much of *Sein und Denken* is concerned with the nature of this impression-of, its genesis, its unity amidst differences (i.e. with the distinction and relation between *the* impression-of-vitality and *an* impression-of-vitality), and its relation to "being". With regard to this last, König takes "Being" (*Sein*) as "to act or to work", and distinguishes, in a manner that is remotely reminiscent of Heidegger, between "being-present-at-hand" (*Vorhandensein*) and "Being" (*Sein*) in the verbal, intensive sense of acting, or rather as so-acting. How are these two senses of "being" and the corresponding predicates — determining in one case and modifying in the other — to be further explicated? Ascribing a determining property ("red") to a thing that is present at hand presupposes that there is already such a thing which is then said to be red. Ascribing a modifying predicate ("sublime") does not presuppose such an already posited entity. The mountain which works majestically is nothing but what works in that manner. Obviously, at this point, any sharp distinction between the two appears to falter, for it would seem to be undeniable that there is a mountain, no matter if it works on some one as majestic or not. Is then being as acting posterior to, and in that sense does it presuppose being as presence-at-hand? König would deny any such conclusion. For him, being as acting is prior. How are we to understand this priority?

König's very difficult view on this matter is something like this: being-present-at-hand (*Vorhandensein*) is not a determining "what" of a thing. Being-a-tree is a determining "what" of *a* tree. Nothing is *a vorhanden.* The tree over there is not *a Seiendes.* The true subject of statements about being-present-at-hand is not this thing over there, but the infinitely many x in locutions of the form "x is present-at-hand". To transform the latter into "x is a being that is present at hand" is misleading. However, in the case of a landscape that works as majestic, or a smile which strikes as mysterious, being and so-working are one and the same. The being is not presupposed in the predication of "majestic". A majestic thing *is,* insofar as it works in a certain manner. König considers this to be true being, one which is not derived from a propositional function "x is present-at-hand". Shall we then say that the so-working is not *vorhanden*? Does it (the majestic landscape) not also exist? Of course, he concedes, being-present-at-hand is a possible predicate of a so-working entity but the latter is more than merely being present-at-hand.

Lest one may object that the so-working entity is merely subjective while the existing thing, the mountain, is objective, König points out that exactly the reverse may indeed be the case: a so-working entity is related to an impression-of-majesty. But a merely given, present-at-hand entity like the mountain over there refers back to a knowing subject from which it is then abstracted and taken to be independent of such knowing.

One way of explicating the originary nature of this intensive sense of being is this: one can deliberately speak of Karl even if Karl is not there or even if Karl does not exist, just because in normal speech about Karl it is presupposed that Karl does exist. There is no such presupposition in speaking of "impression-of-majesty", one therefore does not speak of it when there is no such impression. An impression of majesty guarantees that there is a being which works majestically. Speaking of it is not referring to it through an ideal meaning, one does not speak of it as such and such, but speaks of itself.

We perhaps begin to understand why König calls "impression of" as itself showing its *wovon*. In this sense, he does not mind saying that the impression-of-being is also thinking-of-being. It does not matter to him if we call it thinking or feeling. We feel the innocence of a smile, but this feeling also shows being-innocent, it — shall we also say? — is thinking-of-being-innocent. This is what König calls *Sein-Denken* — not reflective thought, but, as it were, a non-reflective thinking. Being, in this case — being-innocent, being-majestic, being-beautiful — is not a concept, not a feeling, not an impression, but it is the immanent object of an impression-of. The tree over there is not essentially such an immanent object of my thought about it. But being-majestic is so contained — in Brentano's locution, inexists — in the impression-of-being-majestic.[11] Being in this sense is experienced being.

Such are not only impressions-of-beauty, or of-goodness, or of-vitality, but also feeling-of-spiritual-presence, feeling-of-eternity, -of-the-deep-secret-of-existence, -of-the-presence-of-God-in-all-things, -of-temporal-presence, -of-spatial-depth. The point is that it is not as though we already know what it is to be, for example, a spiritual presence and now we have a feeling of it. We rather understand what is being a spiritual presence by having such a feeling. In this sense, the locution "a feeling tells me" is genuine. A genuine impression-of is a speaking feeling in the sense that it alone is the condition of the possibility of the expression which expresses it.

I have tried in the foregoing to present as accurately and also as briefly as I could, the essential distinction that runs through König's *Sein und Denken*. What we need to be clear about is, how this distinction is related to the distinction between practical and theoretical sentences. Theoretical sentences, we learned, are formed out of propositional functions by replacing variables by constants or by quantifying over the variables. The sentence "Mt. Everest is 29,000 feet high" presupposes that Mt. Everest exists, here "existence" has the logical meaning that a propositional function is sometimes or always true. But in "That is majestic", the subject is a that-there as so-working, and we do not have a presupposed function "x is majestic". It would then seem as though König's modifying speech, or sentences in which the predicates are modifying predicates, are practical sentences, and determining speech is theoretical. However, such a coincidence between the two distinctions may be just mistaken. "Socrates is mortal" and "That is my friend Karl" are practical sentences, but they are not examples of modifying speech. Their predicates are rather determining predicates. Let us recall that König recognises two heterogeneous groups of practical sentences: his singular sentences, and the Aristotelian non-apophantic sentences (i.e. sentences expressing requests, commands, etc.). Shall we add a third group to these two: sentences such as "That is sublime"? In that case, some practical sentences would have determining predicates, and some modifying predicates, while others would not be subject-predicate sentences at all. What justifies bringing them all under one genus "practical sentence"? They are all radically different from theoretical sentences, they all refer to a particular context situation or a subject, none of them expresses a propositional thought or an ideal meaning, and those of the first two groups are directly *about* a real entity situation or context. While this seems to justify collecting them all in one group and opposing this group to the theoretical sentences, there is a seeming discordance in such coincidence which must have been noticed. Of theoretical sentences, König says, as we noted earlier, they are themselves (*selber*) true or false, whereas singular sentences like "Karl is blond" are not themselves (*selber*) true or false, for they do not show, of what is the predicate predicated. They state or mean what their utterers state or mean by uttering them. Yet, a sentence such as "That is sublime" expresses an impression-of-being-sublime, and such an impression itself shows its *wovon,* its immanent content. Being so constituted, the sentence "That is sublime" could not but be true. It is not about a presupposed thing, the mountain, for example, but is about the so-working entity, the entity which is working as sublime. Such a sentence is itself true. But we should be able to notice that theoretical sentences are not themselves true, they are themselves true *or* false. "That is sublime", on König's construal, cannot be false.

A singular sentence with a determining predicate refers back to the utterer's intention and context. A singular sentence with a modifying predicate expresses as impression-of, but not any subject's intention. Neither expresses an ideal meaning. Both directly refer to a real being. Neither states that a propositional function is sometimes or always true.

B. In order to better understand König's distinction between the practical and the theoretical, we need, above all, to look into his long essay on the *concept of cause.* This essay is devoted to the theme of developing a distinction between "theoretical cause" and "practical cause". Rejecting the belief — shared by Hume and Kant, despite the otherwise deep differences in their views about causality — that the efficient cause is a thing, an existent entity or an event, localisable in space and time, König warns us against construing "a is the cause of b" as a subsumptive judgment wherein one subsumes an existent, perceived entity *a* under the concept of cause (just as in "a is a dog", one subsumes *a* under the concept of "dog"). On his view, the concept of cause is the concept of a satisfying answer to a why-question. The so-called cause is that which makes us understand why in a satisfactory manner. A why-question, in everyday life, arises in a situation where the normal, expected pattern of action and behavior is upset. Contrary to Hume's well-known thesis, König argues that it is not by repeated observation of the like followed by the like that one arrives at the concept of cause, but that it is when something other than the expected happens that one asks the truly causal question: "why?" It is not when A behaves as it always has, but rather when it behaves differently than before, that we want to know why, and one answers this question by assigning a cause — an efficient cause, more specifically. The answer is not satisfactory because it states what is in fact the cause, but rather what it states is the cause because and insofar as the answer satisfies us in the sense that it restores for us the lost secure ground, the ground which the unexpected behavior was in need of. The cause then, on König's view, is expressed by a thought-structure, patterned in a definite manner: "this billiard ball there which was at rest is now in motion, *because* it has just received a thrust and because a ball which while at rest receives a thrust is a ball which begins to move."

With this much of introduction to König's general view regarding causality we can introduce the distinction between practical cause and theoretical cause. A practical cause is what is given in a satisfying answer to a practical why-question concerning a practical this-here-and-now: why does *this* ball over there which was at rest, now begin to move? A practical why-question is a question of this sort. What is called an efficient cause is, for König, really a practical cause: one asks the practical why-question initially out of the pressure of action.[12] As contrasted with it, a theoretical cause is what is stated in a satisfying answer to the question of the sort, why is it that a ball which has received a thrust is a ball in motion, i.e. why does this general implication hold good? In the case of the example on hand, the theoretical why question amounts to asking, why is it that the laws of impact are what they are? Answering such a question is to make a typically metaphysical move — as is illustrated in the thoughts of Leibniz, Goethe and Schopenhauer, philosophers who deeply interested König.

Since, as should be clear even from this cursory account of his extensive discussion, König rejects the distinction between real ground and epistemic ground and, as a matter of fact, identifies the cause with ground, or rather with premises, the effect, even the practical effect, is construed by him as the *content* of a factual alteration of a practical this[13] — as a content that is understood as deducible from certain premises. Consequently, it is strictly speaking meaningless, on his view, to regard effects as alterations. The *content* of an alteration is what one *says* when one says which alteration it is: thus, the content logically refers back to men, while this is not true of the alteration itself.

C. The distinction between an alteration and its content helps us to understand König's central point of criticism against Nicolai Hartmann's theorem regarding the necessity of all real events.[14] Hartmann, as is well known, argued that since real-actuality (Realwirklichkeit) implies real-possibility (Realmöglichkeit), and since real-possibility of an event is nothing but the completion of the series of real causal conditions that bring about that event, it follows that the real-actuality of an event implies its real-necessity as well. In other words, precisely that which makes an event really possible, also makes it really necessary. König not only finds this argument flawed, but uses the occasion to question the very conception of ontology that Hartmann operates with.

A real entity in Hartmann's terminology is a real-actuality, it is something that is not repeatable. B is said to be a condition of A, if B is always there in case A is, and A sometimes is, if B is there. Now can we meaningfully use the adverbs "always" and "sometimes" of A, if A, as a real actuality is an unrepeatable *hic et nunc*? Hartmann so defines "actuality", "ens" or "Seiende" that actual existence, *Dasein* or *Vorhandensein* belongs to its essence — which is in sharp contrast to the traditional ontological understanding that an entity is something which can possibly exist, and therefore may not possibly exist. It is obvious then that when one is talking, in Hartmann's sense, of a real entity one is not talking about a stone or about an event, but about *this* stone and *this* event. Let us now recall König's distinction between practical *this* and theoretical *this*: a practical this is a pure this, this this-there for me, to which it is accidental what it is, a book or a stone. But a theoretical this is a this such and such — this man or one such as this. Real existence belongs to the essence of a practical this. Practical this and Hartmann's real actuality are identical. What Hartmann calls "real conditions" are conditions, not of this entity over there i.e. of the practical this as such, but rather of an entity that is so constituted, of any such entity — in other words, of a theoretical this. The practical this, as a matter of fact, is a limit to theory. The chain or totality of such practical thises cannot constitute a closed world. It cannot constitute a closed world, just because it will be ever open-ended, and secondly because a practical this is essentially relative to some one. If the world were a practical this for some one, this some one, on its part, would be a practical this for some one else.[15]

It would appear as though part of what König says in his critique of Hartmann is inconsistent with his thesis in the essay on causality. In the former, a practical this is characterised almost as if it is *causa sui* (for all that may be regarded as its causal conditions would really belong to its constitution. That stone rolling down the hill, as a practical this, is not caused by that strong wind, but it is rather that-stone-rolling-down-that-hill-under-the-impact-of-that-strong-wind which is the real event. One can thus go on indefinitely enriching the internal content of this event.) In the essay on causality König characterises the practical cause as the cause of a practical this in the sense of providing a satisfactory answer to a practical why-question. As a matter of fact, there is no real inconsistency between the two. The former thesis, namely that a practical this is *causa sui* is only a consequence of a position like Hartmann's. It is not a position that König himself accepts, for on his view a cause is not a real event, it is rather a thought-structure.

Let me now try to state perspicuously the ontological distinction which underlies all this. There are two kinds of entities: *things* in the ordinary sense, such as sticks and stones, and entities referred to by practical "this"es, subjects of modifying predicates such as "sublime". The former i.e. things in the ordinary sense are *ens* in the sense of classical ontology. For them, to exist is as much a possibility as not to exist. The latter *are* in the verbal intensive sense, existence belongs to their very essence. The former sort of entities constitute a closed world; the latter constitute the lived everyday experience. Our natural why-questions arise in the context of the latter; our satisfactory answers to such questions involve use of general implications which make use of the former. The search for theoretical causes takes us beyond the latter discourse to speculative discourse about "God's infinite wisdom and goodness" (Leibniz), or the Universal Will (Schopenhauer) or the speculative concept (Hegel).

III

A. A sentence in which a practical "this" occurs, is, according to König, *eo ipso,* a practical sentence.[16] But, as said before, sentences containing a practical "this" are not all of the same kind. These sentences may differ, first, with regard to the nature of the predicate (the predicate may be either determining or modifying) and, secondly, with regard to the subject which may be either a *pure* practical "this", or a practical "this" to which some descriptive words are attached, or an ordinary (neither Husserlian nor Fregean) proper name such as "Karl". Thus we have:

1. This is red.
1*. This is majestic.
2. This stone over there is rolling down the hill.
2*. This landscape is sublime.
3. Karl is blond.
3*. Karl looks serene.

The distinction between pure practical this and a practical this which yet contains a descriptive element is not explicitly made by König. But it appears, he chooses examples of practical sentences from both groups i.e. from those whose subject is pure "this" and from those whose subject term is "this" together with some descriptive phrases. With regard to 2 and 2*, it is important to be clear as to how they differ from the theoretical sentences containing theoretical "this". A theoretical "this", we are told, means one such or a thing such as this.'[17] At this point of the discussion, I must confess I have not found in König's writings one single example of "this" in which it is used as a theoretical "this". What we have are locutions like "a thing such as this". What is the word "this" occurring in this phrase? Is it a theoretical or a practical "this"? I would think, there are only two possible construals. One may regard the "this" (as occurring in "a thing such as this") as a practical "this" and hold that it is the modifier "a thing such as" which makes, not the "this", but the entire expression "a thing such as this" into a theoretical expression, which only shows that a practical expression — not only "this", but also an ordinary proper name such as "Karl" — can be a constituent of a theoretical expression. Under this construal, there is no theoretical "this", all use of this demonstrative is practical in the sense explicated earlier. The other possible construal is to regard "a thing such as this" as one composite expression ("a-thing-such-as-this") and to treat it as theoretical and as an inseparable component of it, to treat the "this" occurring in it as theoretical. I favour the first construal as being simpler, and therefore I find König's idea of a theoretical "this" suspicious and would prefer to regard all use of "this" *alone* as practical (while conceding the possibility of using it as part of a theoretical expression). But these remarks do not affect the validity of König's distinction between the theoretical and practical sentences.

B. Let me now turn to the case of proper names such as "Karl". Clearly, König would reject what is called the description theory or attribute theory of proper names — whether in the original Fregean form or in the form it has been given by John Searle. Nor will he accept the now fashionable direct reference theory or some version of the causal theory made popular by Kripke amongst others. To me, König's theory of proper names seems to contain elements of an account developed by Tyler Burge[18] and elements of a theory developed by Castañeda.[19] Like Burge, König holds that a proper name really contains (covertly) a demonstrative element. For him, "Caius" is a practical this.[20] According to Burge, "Karl"≡"This Karl". On both the theories, demonstrative reference belongs to speech acts, not to a language system. König's practical this, or his practical sentence, is an utterance. The written sentence "That is my friend Karl" (given as an example) is only potentially practical.[21] Beyond this, of course, the two theories diverge, and diverge radically. On Castañeda's theory, there is a semantic openness about proper names. "Karl" may be the name of my friend, or of my neighbor, or of innumerable others, in which case the uniqueness of its reference is due to the speaker's intention. The proper name, then, functions in the speaker's *idiolect,* and its reference cannot be determined externally.[22] Furthermore, according to Castañeda, the singularity of reference of a proper name belongs, not to that name by itself, but to the entire speech act, the whole uttered sentence along with its context. Castañeda also recognises that indexical reference is perforce personal and "essentially ephemeral",[23] besides being confrontational and practical (in his language executive).[24] It is surprising how close these characterisations are to some of König's. Let me recall them for the present purpose.

Arguing for his thesis that a practical sentence is not itself (*selber*) true or false, König insists that when one hears a speaker say "Karl is blond" and understands it, one still does not know whom the speaker is talking about, and so cannot even ask, is it true? An auditor may say Karl is a person and is the speaker's neighbor; another may say she knows that Karl is the name of a dog. König's point is that there is no non-circular way of stipulating who Karl is. As a matter of fact, on König's view, the word "Karl" is not "immediately" meaningful, whereas the word "man" is. A sentence containing "Karl" has to be interpreted in the context of a definite occasion. Demonstrative reference — in König's locution, the practical this — is relative to a subject, it is a this for someone.[25] What he calls the "practical truth" is momentary (*augenblicklich*).[26]

In spite of these agreements, König's theory and Castañeda's theory differ widely in other respects, two of which may only be mentioned here. In the first place, König believes that whereas theoretical reference is *via* a (Fregean) ideal meaning, a practical sentence has not such ideal, propositional meaning, but directly refers to an entity's being so and so. From the point of view of communication, what is communicated by the utterance of a theoretical sentence is a proposition, while what is communicated by a practical sentence is a being, e.g. Karl's being blond. Let us recall that on his view asserting a theoretical proposition is asserting that a propositional function is sometimes or always true, while asserting a practical sentence is expressing practical knowledge of an entity itself. Castañeda's theory has just the opposite implications. In general, he holds that even demonstrative reference is mediated by cognitive contents, and along with Burge regards sentences containing names of individuals as expressing propositional functions. "A name has the logical role of a free variable of quantification, indicating the position of an element of the propositon before the mind of the speaker, which element he is leaving unexpressed."[27] This is as far from König's view as possible. The issue is deep and needs further exploration.

C. The practical sentences which most figure in König's writings belong to types represented by 2 and 2* — the former sort in the essay on causality and the latter in *Sein und Denken* as well as in the inaugural lecture on aesthetic effect. Let us therefore look at them again. In "This stone over there is rolling down the hill" (or, more accurately, "This stone over there which was at rest now begins to roll down the hill"), the subject term "This stone over there" does not have the semantic openness of the pure "this", it does not also have the semantic openness of "Karl". Knowing the context, i.e. the occasion of the utterance is of course necessary to remove whatever openness it still has. We know, from the mere understanding of the sentence, that this-there is a stone, not a billiard ball. We need to know the speaker's situation in order to be able to identify which stone, where and when. In this respect, it is like "This is Karl", and may indeed be subsumed under "practical sentence". The same may be said of "This landscape is sublime", although the modifying predicate "sublime" requires different treatment than the determining predicate "blond" or "called Karl".

D. König's works fall on the borderline between logic, ontology and philosophy of language. I think, while there is an ontological position underlying all this, that ontological position is made subservient to a logical thesis which again is established on the basis of a close look at ordinary language use (sometimes also, as in *Sein und Denken,* on the basis of a Heideggerean sort of etymologising.) As far as the distinction between the theoretical and the practical is concerned, the historical roots reach back, as mentioned at the beginning of this essay, to Aristotle, but, in more recent past, align themselves with Dilthey, Misch and Heidegger (Husserl remaining a constant presence in the background). Dilthey is present through Misch (König's immediate teacher), and what is particularly influential is Misch's distinction between *Lebensverständnis* and objective cognition, or, correspondingly, between hermeneutic statements and logical judgments. The connection with Misch is explicitly referred to, and its close proximity acknowledged in the essay intended for his *Festschrift.*[28] As far as Heidegger is concerned, the well-known distinction between the two modes of being — being as presence-at-hand (*Vorhandensein*) and being as being-for-use (*Zuhandenheit*) — is recognised as an antecedent, as also the distinction between "hermeneutic as" and "apophantic as" (which is also close to the Misch distinction mentioned above). Let me now briefly expand on these connections.

In one of his *Entwürfe* in theory of knowledge and logic, of around 1880,[29] Dilthey insists that the categories of presentday logic are to be distinguished from the categories and concepts of life, and that the former are grounded in the latter. As an illustration, he says that the logical concept of predication is grounded upon a *Lebensbegriff* of selfsameness (*Selbigkeit*). Around 1893, he writes that thinking is a function of life.[30] Such concepts as "thing", "effect", "suffering", "substance" are basically categories of life. The categories of "essence", "Bedeutung" and "Sinn" are abstractions from the "Zentrum der Lebensstruktur selbst".[31] Life has a silent knowledge of itself, which Dilthey even calls, at least sometimes, "Denken", and which is the basis of the discursive thinking of logic. Misch often speaks of a logic which corresponds to the intention of *Lebensphilosophie.*[32] Such a logic will comprehend the two aspects of life: its *Gedankenmäßigkeit* and its *Unergründlichkeit.* In an appendix to his essay on Misch, König reports that Misch thought of his own logic as an extension of transcendental logic from within to the objects of the *Geisteswissenschaften* on the basis of one's original understanding of life. He also calls it hermeneutic logic and contrasts it with the apophantic logic of discursive thinking. We also learn from König that Misch spoke of discursive logic as being concerned with the "ontic", by which he meant everything that is given to the cogniser in outer perception.[33] Discursive statements can be, in principle, confirmed or disconfirmed by outer perception. Hermeneutic logic is the logic of our "inner" understanding of life, of values and goods, of the meaning and significance of situations. Misch also calls it "practical knowledge".[34] Hermeneutic logic is a mode of interpretation of life. The interpreter, in such a logic, is not an individual thinker, but language itself, or as König puts it, "*der sprechende Mensch als solcher*". Consequently, this logic is not the result of a reflection on life, but rather a product of life's self-interpretation. Thus, whereas the concept of intentionality belongs to discursive thinking, the concept of "*Lebensbezug*" belongs to hermeneutic logic.[35]

König concedes[36] that his own distinction between theoretical and practical sentences is essentially the same as the distinction between judgments in the narrow sense and hermeneutic judgments, but he also adds that the two distinctions do not coincide. In his own formulation, the logical comes to the forefront, and the aspect of *Lebensphilosophie* recedes to the background.[37]

Similar remarks may be made in connection with the Heideggerean distinction. Practical being is being ready-for-use (*Zuhandensein*), and theoretical entities are just present-at-hand (*vorhanden*). While this is what König says, I must insist that not all König's practical this-es are *zuhanden* in the same sense of being ready-for-use (as a hammer is, to recall Heidegger's example). When he says "That is my friend Karl" is a practical sentence, it is not because my friend Karl has a mode of being which is best understood as *Zuhandensein*, but because these sentences — not merely their utterances — are themselves, in a rather peculiar sense, actions[38], they are actions in a sense that is closer to, but not identical with, the sense in which "I promise" is itself an action. They are addressed to some one, and this being addressed, and the occasion of that address, are parts of their meaning.[39]

However, after recognising the proximity of his own distinction to Heidegger's, König goes on to add that on his view, the ontological distinction between the two modes of being is grounded in, rather than grounding, the logical distinction between the two kinds of sentences.[40] It is only because one generally accepts the Aristotelian thesis that there is no logical distinction between the two sorts of sentences, that what one sees underlying their difference is taken to be an ontological distinction, whereas in truth a distinction which is fundamentally logical gives rise to the impression of an ontological distinction. In other words, the truth of the ontological distinction is in the logical distinction.

Even if we accept this last claim, one cannot but ask, what sort of logic is it to which this distinction belongs? Not to formal logic — no matter, whether Aristotelian or mathematical, whose domain is theoretical discourse; not to hermeneutic logic whose domain is practical discourse. It has to be a philosophical logic which comprehends them in their distinctness as well as in their interrelation.

E. I said at the beginning of the preceding section (D) that there is an ontological position underlying König's distinction, although he would let ontology be in the background of his philosophical logic. The ontology emerges nowhere in a clearer light than in his discussion of the concept of truth. If the two kinds of sentences are radically different, then the senses in which they are true or false must also be radically different. This indeed is the case according to König. "Theoretical truth" is logically different — and radically different — from "practical truth". If the concept of being is logically connected with the concept of truth (as e.g. Frege held in his thesis that to say there are numbers is equivalent to saying that number-statements are sometimes true), then we would be having on hand a distinction between "theoretical being" and "practical being". Let us quickly, and also briefly, look at both. A theoretical sentence is true if the corresponding state of affairs obtains, and the state of affairs obtains if the theoretical sentence stating it is true. A practical sentence is true — and this is the original, and also likely to be the most controversial part of the theory — if some one, using it, speaks the truth, and one speaks the truth if one says what one "practically" knows. Thus to vary upon König's example, if I "practically" know my brother is now in Chicago (whereas as a matter of fact, my brother, unbeknownst to me, is in Seattle), my saying my brother is in Chicago is saying the truth. It is not that I suppose my brother is in Chicago (which I may be doing in some other case). If I am supposing, guessing, imagining, then I do not have the "practical" knowledge. But when I know my brother is in Chicago, I do not know that a certain proposition is true, I know a practical being, in this case, my brother's being in Chicago. The opposite of "practical" truth saying is lying, not error. In order to interpret König correctly at this point, it is important to note that his concept of "practical knowing" is such that it is at least possible to ascribe practical knowledge to animals.[41] My dog possibly knows ("practically", and without entertaining a proposition) his bone's being under the table. It is also important to recall that König totally rejects a likely interpretation of his thesis to the effect that practical knowledge is subjective, while theoretical knowledge is objective. Underlying this likely misconstrual there is again the mistaken belief that practical and theoretical sentences are logically of the same sort and that consequently their truth-character is the same. In other words, one wants to say that my practical knowledge that my brother is now in Chicago is only subjectively true (in case he is really in Seattle), for it does not correspond to the objective state of affairs. But that amounts to denying the deep difference that separates the practical from the theoretical. If we want to preserve the radical difference between the two, we should rather say that in in practical knowing has its own sort of objective truth, which should not be made to coincide with the objectivity of theoretical knowledge. It is true that practical knowing, truth and being have a reference to a subject. But they are not for that reason subjective. Let me quote König at this point:

> Die Subjektivierung des praktischen Seins hat zugleich die Bedeutung einer Objektivierung des praktischen Wissens: das praktische Wissen ist kein zu-wissen-Meinen. Und entsprechend umgekehrt: die Objektivierung des theoretischen Seins ist gleichbedeutend mit einer Subjektivierung des theoretischen Wissens als *Idee* eines wahren Meinens.[42]

Practical being is in one sense subjective — in the sense of being for some one. It is also strictly objective, for one who knows "practically" knows a practical being itself, and not a propositon. A practical being is such that it is, essentially a known being. This is not true of theoretical being.

The above entails that different subjects may have conflicting practical knowledge, so that there may be simultaneously conflicting practical beings (my-brother's-now-being-in-Chicago and my-brother's-now-being-in-Seattle). The idea of practical knowing does not require that these conflicting practical beings must all converge towards the truth. However, that precisely is the idea of theoretical truth: it is an idea of adequacy which men aim at. But practical truth, as truth-saying, is not an idea. One could express the contrast in another way: theoretical being is only meant, intended being, as the Idea of perfect adequacy to be aimed at, while practical being is being itself, being which is essentially known as such, not pursued as an Idea.

What I called an underlying ontology is now clear. Being itself — and not merely a thought, an Idea — is known "practically". In discursive, theoretical thinking, one pursues an infinite goal of an Idea, but never quite reaches being itself. Some version or other of such a philosophical view can be found in many philosophers. König's is an almost unique version of that general philosophical belief. The priority it gives to logic over ontology is misleading, for the truth-character of theoretical discourse (with which formal logic is concerned) is devalued as an Idea, whereas practical knowledge is linked directly to being itself (being itself being at the same time the correlate of practical truthfulness). The radical distinction between them is a *logical* distinction in a rather philosophical-logical sense. It would perhaps be more appropriate to say that what König achieves is the correlation between ontology, logic and cognition — not a priority of anyone of the three.

After all this is said and put in the proper perspective, there remains the enormously difficult problem of how to understand the complicated manner the two — the theoretical and the practical — are interrelated, in logic, language, knowledge and the world. König was fully aware of this problem, and devoted much of his thinking to it. I reserve a discussion of his views on this issue for another occasion.

NOTES

[1] Husserl, E., Ideen zu einer reinen Phänomenologie und phänomenologischen Philosophie, Erstes Buch, ed. W. Biemel, The Hague, 1952, § 13.

[2] König, J., *Der Begriff der Intuition*, Halle, 1926.

[3] Quoted by König in his lecture on aesthetic effect (see ref. under 4 below).

[4] König, J., *Die Natur der ästhetischen Wirkung*, originally appeared in K. Ziegler (ed.), *Wesen und Wirklichkeit des Menschen*, Festschrift für Helmuth Plessner. Göttingen 1957 and now reprinted in G. Patzig (ed.): *J. König. Vorträge und Aufsätze*. München/Freiburg 1978. The references will be to this volume under the abbreviation *VuA*.

[5] By "logic lectures", I am referring to König's two-semester long lecture course on *Theoretische und Praktische Sätze*, given in Göttingen during the summer semester 1953 and winter semester 1953/54. I make use of my own lecture notes.

[6] I am indebted to Professor F. Rodi for making a copy of this essay entitled *Der Logische Unterschied theoretische und praktische Sätze*, available to me. I will refer to this essay as *TuP*. The paginations are from the printed galleys. The essay was never published.

[7] Lukasiewicz, J., *Aristotle's Syllogistic from the Standpoint of Modern Logic*, Oxford, 1951, 1-7.

[8] König, J., *Sein und Denken*, Studien im Grenzgebiet von Logik, Ontologie und Sprachphilosophie, Halle, 1937.

[9] König, J, *Bemerkungen über den Begriff der Ursache*, originally in: *Das Problem der Gesetzlichkeit*, ed. by Jungius Gesellschaft, Hamburg, 1949, Reprinted in *VuA*. All references will be to this volume.

[10] See note 4 above.

[11] *Sein und Denken*, 93-94.

[12] *VuA*, 144-5.

[13] *VuA*, 252.

[14] For Hartmann's argument, see his *Möglichkeit und Wirklichkeit*, Berlin, 1938. König's essay on Hartmann is entitled *Über einen neuen ontologischen Beweis der Satzes von der Notwendigkeit alles Geschehens*, in: Archiv für Philosophie 2 (1948), 5-43. Reprinted in *VuA*.

[15] *VuA*, 111.

[16] *VuA*, 154fn.

[17] Ibid.

[18] Burge, T., "Russell's Problem and Intentional Identity," in: James E. Tomberlin (ed.), *Agent, Language, and the Structure of the World*. Indianapolis, 1983.

[19] Castañeda, Hector-Neri, *Thinking, Language, Experience*, Minneapolis, 1989.

[20] *TuP*, 162.

[21] *TuP*, 155-6.

[22] Castañeda, loc. cit., 30.

[23] Ibid, 69.

[24] Ibid, 70.

[25] *TuP*, 62.

[26] *TuP*, 157.

[27] Castañeda, loc. cit., 251.

[28] *TuP*, 139f.

[29] Dilthey, W., *Gesammelte Schriften* XIX, Göttingen, 1982. The fragment is called *Frühe Entwürfe zur Erkenntnistheorie und Logik*.

[30] Ibid, 347.

[31] Ibid, 375.

[32] König, J., *Georg Misch als Philosoph*, Göttingen, 1967, 208.

[33] Ibid, 224.

[34] Misch speaks of this knowledge as "in seinem Vollzug zugleich 'praktisch', für das Leben verbindlich und in es eingreifend." (Misch, *Lebensphilosophie und Phänomenologie*. Eine Auseinandersetzung mit Heidegger und Husserl. Leipzig/Berlin, 1930, 233.) Cp. König's remarks in this connection in *Georg Misch als Philosoph*, 225.

[35] *Georg Misch als Philosoph*, 230.

[36] *TuP*, 137.

[37] *TuP*, 141.

[38] *TuP*, 119.

[39] Klaus Hartmann's label "judgment of address" seems appropriate. I found this in J. Claude Evans, *The Metaphysics of Transcendental Subjectivity. Descartes, Kant and W. Sellars*, Amsterdam, 1984, 63.

[40] *TuP*, 143.

[41] *TuP*, 153.

[42] *TuP*, 161.

8. LASK'S THEORY OF JUDGMENT

(for Thomas Seebohm)

Lask's theory of judgment has to be understood in the light of two other theories: the Kantian and the more recent Bolzano–Husserl theories. Let me therefore begin with a brief statement of these two theories.

A. *Kant's Theory*: Kant does not clearly distinguish between the act of judging and the product of this act, i.e., the judgment. Some of his definitions apply to the one, some to the other, as Allison has beautifully shown.[1] In any case, judgment cannot be defined as the representation of the relation between two concepts. The crucial question is: what sort of relation between two concepts is being represented in judgment? Kant's answer seems to be that the relation, which is one of synthesis, must be *objectively valid*. In other words, the concepts being synthesized must not be connected merely subjectively, e.g., according to the laws of association. The synthesis must be in accordance with an objective rule, which is nothing but one of the categories of the understanding.

Besides this idea of synthesis in accordance with an objective rule (whereby a judgment acquires objective validity, i.e., is either true or false), Kant's theory has three other important features (leaving aside, for my present purpose, the analytic-synthetic distinction). In the first place, Kant still required, in consonance with pre-Kantian views on the matter, that a judgment must have at least two concepts, one of which is the subject and the other the predicate. In an important sense, both are, for Kant, predicates. The one which directly relates to the object (being judged about) is regarded as the subject; the one which indirectly relates to it, i.e., through the mediation of the first, is the predicate. Thus the judgment 'A body is divisible' is equivalent to "There is an X which is a body and is also divisible".

Secondly, Kantian thought rejects the absolute distinction drawn, in pre-Kantian thinking, between concept and judgment. A concept is not only a possible predicate of a judgment, as Kant tells us, but involves an "analytic unity" whereby raw representations are *interpreted* as belonging under one representation. But analytic unity and synthetic unity go hand in hand,[2] so that a concept already involves judgments. There is no pre-judgmental level of thinking, that amounts to a mere entertainment of a concept. Having a concept involves the ability to make appropriate judgments under appropriate circumstances.

Thirdly — and this concerns not the structure of judgment but its epistemic value — we owe to Kant the insight that to know is to judge. At the same time, the Copernican revolution requires that the object of knowledge is constituted precisely by the very same rules of synthesis which also constitute judgment. The object, then, does not belong to a metaphysical domain outside of the domain of logic (conceived not as formal, but as 'transcendental' logic) — as it was for pre-Kantian modes of thinking, but — in so far as its formal constitution is concerned — precisely belongs to transcendental logic.

B. *The Bolzano-Husserl Theory*: This Kantian theory of judgment stands in sharp contrast to a theory which can be traced back to Bolzano and had two distinguished proponents in Frege and Husserl. For my present purpose, I will only recall this theory in its general features without dwelling on the differences between the ways its various proponents developed it. The theory makes a distinction — which the Kantian theory did not — between the act of judging and the judgment as an ideal (i.e., not spatio-temporally individuated) objective entity (i.e., intersubjectively available and independent of any particular judging subject's mental life). The act of judging then consists in recognition of the truth-value of this ideal content (Frege), or in accepting or rejecting it (Brentano), or in affirming or denying it. Mental acts, qualitatively different from judging, may also be directed towards the same content: one may doubt, assume, suppose or question the same content, as well as affirm or deny it. The mental acts belong to the real, mental lives of human thinkers, and so belong to the domain of psychology; the ideal — objective contents — also called 'thoughts' or 'propositions' — form the proper subject matter of logic. A theory of judgment, on this account, should concern itself with the logical structure of the latter. On a rather influential version of this structure, a simple thought or proposition consists in the *sense* of a name and the *sense* of the predicate expression. Its reference consists in an object (which is the referent of the name which figures in the sentence) and the concept which is the referent of the predicate expression: these two elements, taken together, amount to either the True or the False — the two truth values the thought may have. On another account, the propositional content is an unanalyzable whole which is simply accepted or rejected in the act of judging.

How does this theory of judgment relate to the pre-Copernican mode of thinking (as with Brentano) or can form part of a Copernican mode of thinking (as with Husserl)? In the former case, one has to determine how the propositional content (that is accepted or rejected) is related to the things that are out there in the world; in the latter case, the thought contents, ideal meanings, have to play a role in the constitution of the world.

C. *The Theory of Geltung*: There is however a version of the theory of ideal contents which, first formulated by Lotze, reappeared within the late nineteenth century Neo-Kantian tradition. Thus the two sorts of theories of judgment appear together within this version. Lotze characterized the mode of being of such ideal contents, propositions or thoughts to be one of *Geltung* or validity (they are either true or false, 'true' and 'false' being *valuational* concepts), and distinguished it from the mode of being of real things which simply are, i.e., have being or *Sein*. Logic is then a normative science. Its domain is defined by the concept of *Geltung* or *Sollen*. Kantian transcendental logic is concerned with the a priori forms of objective validity, of those *Geltungs*-forms which determine the domain of objects. Thus the concept of *Geltung* comes to have a priority — in accordance with the Copernican revolution in philosophy — over the domain of *Sein*.

Emil Lask's importance lies precisely in the fact that he, of all philosophers of his time, stands at the meeting point of these *three* modes of thinking: the Kantian, the Bolzano-Husserl and the Lotze-Neo-Kantian. He appropriates elements from all three, and goes beyond them to construct a highly sophisticated theory of his own.

II

Lask's Relation to these Theories

One of Kant's decisive contributions to the philosophy of logic was to have drawn the distinction between formal logic and transcendental logic. Since this distinction was of fundamental importance for Lask, let me pause for a moment to dwell on this Kantian thought. Kant's own way of drawing the distinction[3] is that while formal logic abstracts from all contents of thought, transcendental logic abstracts only from empirical contents — which implies that transcendental logic also takes into account the pure contents of thought, i.e., the pure intuitions of space and time. But space and time are bare *forms* of intuition, and belong to the subject matter not of logic but of aesthetic. Logic's — i.e., transcendental logic's — concern is with how the forms of thought relate to possible objects of intuition (in space and time) — in other words with the categories (and not with the mere forms of judgment). The categories are the forms, not of empty thought as such, but of thought that is objectively *valid* (where 'validity' should not be construed as 'truth'[4]). Thus one may say that transcendental logic yields the a priori principles necessary for objectively valid (i.e., either true or false) thinking, while formal logic yields the principles of merely consistent thinking with no reference whatsoever to possible objects of thought. As has already been indicated, the idea of *validity* was soon regarded as being close to that of *value*. The decisive step that led from Kant to the Neo-Kantianism of Heidelberg school was mediated by Lotze's insistence that it is only sentence-meanings, propositions — and not word-meanings or concepts — that have ideal being. As long as one took the independence of logic from psychology to warrant the existence of word meanings as ideal entities (as Husserl tended to do), one did not come to see that logic was in reality a valuational science. Thus writes Rickert of his own position:

> Ich habe versucht, zu zeigen, warum nur die Bedeutung von *Sätzen* zugrunde gelegt werden darf, wenn man das logische verstehen will, und sobald man den Sinn wahrer Sätze analysiert, kann der Wertcharakter der logischen Gebilde und die Unmöglichkeit einer ontologischen Logik auf die Dauer nicht verborgen bleiben.[5]

In this quotation, Rickert says: as soon as one analyses the meaning of the *true* propositions, one comes to recognize the valuational character of logical entities. But, why true propositions alone, if transcendental logic is concerned with objectively valid propositions, and validity is not truth, but rather truth-or-falsity? This is the point where one can pass over to Emil Lask, who starts from what he calls the value-oppositional character of judgments. For Rickert, value is absolutely worthy of being affirmed ("absolute *Bejahungswürdigkeit*").[6] Lask rejects this.[7]

A. Lask's criticism of Kant:

Kant's great contribution, according to Lask, lay in distinguishing between the two levels of logic — formal and transcendental — the former being non-objective, the latter objective or "material" logic, and in making the former secondary and derived in relation to the latter. Accordingly, according to Lask, it is a mistake to regard judgment as the ultimate subject matter of logic. Within the Kantian mode of thinking, judgment acquires a lower status in relation to the objective categories. The pre-Kantian tendency to regard the objective logic to be also concerned with judgment, despite the Copernican revolution, has to be overcome — and this is the task Lask sets for himself.[8] For pre-Copernican logic the objective domain belongs to metaphysics and provides the original (*Urbild*) of which the judgment, defining the sphere of logic, is the copy. The Copernican revolution upsets this picture by bringing the orginary object within the scope of the logical, i.e., by insisting that the logical is not restricted to the secondary sphere of judgment. The originary is not meta-logical. However, it would be a misinterpretation of the Copernican revolution to say that the object is now shown to be constituted by the *judgmental* logic — a familiar (mis)interpretation of Kant.[9]

According to Lask, Kantian thinking missed the full implications of the great discovery on which it is founded: it, amongst other things, tended to identify "validity" with positive being.[10] It did not explicitly realize that even the negative, invalid synthesis represents relations between matter and categorial relation. Connected with this failure to distinguish between positive value (as opposed to negative disvalue) and being in a sense that goes beyond the opposition between the positive and the negative, is the way Kant expressly connects the transcendental unity of apperception to the copula (which is a judgmental structure). Equally misleading is Kant's failure to realize that positivity and negativity (which are oppositions characterizing only the judgmental domain) cannot have the status of objective categories.[11] In fine, Kantian thinking tended to raise the judgmental structure to a transcendental = constitutive status — contrary to its own insight that the formal logic of judgments must be grounded in a transcendental logic of objectivity. That is why, in spite of the Copernican revolution, judgment continues to enjoy the highest status within the domain of logic — as a consequence of which the "distance" of judgment from the objective-logical is "covered up".[12] Lask's aim is to expel judgment from the domain of transcendental logic, and at the same time to insert it back into the larger structure of the domain of logic as extended by Kant. Judgment lacks any "constitutive" role, it has no transcendental and objective significance; yet it is somehow directed towards the object (from which it is nevertheless removed by an unbridgeable gap). Lask wants to show that owing to its unavoidable "distance" from the object, judgment "modifies" the object into an "artificial" structure; and this "artificial structure" points beyond the judgmental domain.[13]

B. Lask on the Bolzano-Husserl Theory:

The Bolzano-Husserl theory emphasizes the role of *Sinn* in judgment. The judgmental act is directed towards this meaning content which, as it were, occupies a middle position between the object and the judgmental act and which has a certain transcendence with regard to the subjective act. Not unlike Husserl, Lask wants to maintain that the meaning structures come into being through "die fatale Aktivität des Erlebens"[14] and yet, as it were, "float" before the same subjectivity with a sort of quasi-transcendence. However he is also aware that the talk of creation of "meanings" (*Sinne*) is misleading. What subjectivity does is to *separate* the component elements of the original object and lend these separate elements a sort of independence; the *Sinne* consist of putting these separated elements back together (164/5). Once these isolated "artificial" elements are set up, there would exist between them — harmonious or disharmonious — relations in timeless eternity.[15] Thus the *Sinn* is neither wholly subjective nor wholly objective.[16]

While Husserl came to recognize both this subject-dependence and intersubjective availability of *Sinne*, his account of their "origin" is quite different from Lask's. I will not deal with this difference in this paper. For the present, let us note Lask's real objection against all theories of pure logic of *Sinne* — which would include Bolzano's and Husserl's (and perhaps also Frege's).

Bolzano's "truths-in-themselves" and "falsities-in-themselves", as well as Husserl's ideal sentence-meanings are, according to Lask, nothing but the correct/incorrect *Sinne* separable from the acts of judgmental decision.[17] Lask also says that the *Sinne* that Bolzano and Husserl recognize are "oppositionally split", not in the originary, opposition-free *Sinn*.[18] Thus all traditional theories of pure logic of *Sinne* fail to recognize pure transcendence. They only know of a quasi-transcendent *Sinn* which is only made loose from subjectivity but not totally independent of it.[19] Let us try to understand Lask's contentions.

III

The Sinn, *the* Sinn-*fragment and completed judgment*

(i) Consider the judgment 'a is the cause of b' (where 'a' and 'b' are names, not variables). In this judgment, we must first — following Brentano — separate out the merely represented content, what is simply the object of *Vorstellung*. This much should be common to affirmation and negation: in each case, the same content is grasped and then either affirmed or denied. In interrogation, the same content is grasped, but no decision is made about it.

What is this merely represented content? Let it consist in the mere coupling of the elements 'a', 'b' and 'cause' and be written as <a, b, cause>. This is only a *Sinn*-fragment. Both affirmation and negation are about "the same": this fragment. So far, Lask agrees with Brentano. The copula belongs to this fragment.

(ii) From this fragment, Lask distinguishes the full *Sinn* of the judgment "a is the cause of b". It consists of the fragment <a, b, cause> and the value-quality of truth (*Wahrheitsgemässtheit*). The full *Sinn*, then, = <the *Sinn* fragment, value-quality>.

Affirmation and negation are linguistic expressions of the ascriptions of the value and disvalue qualities respectively to the fragment.

(iii) When one simply understands a full judgmental *Sinn*, but does not make a judgment, the value-quality is simply "represented", but not ascribed to the *Sinn*-fragment. *This* represented content looks like: [<a, b, cause>, truth] and is a *Sinn*-fragment of a second order. To say that the judgment is true is to say that this *total Sinn* is harmonious, i.e., there is harmony between <a, b, cause> and the value-quality ascribed to it. These harmonious structures (or, disharmonious structures) may exist timelessly. There are eternal truths and falsities in this sense. These are Bolzano's truths and falsities in themselves.

Lask uses the words "*Wahrheitsgemässtheit*" (and its opposite "*Wahrheitswidrigkeit*") and "*Richtigkeit*" (and its opposite "*Falschheit*") to stand for different things. The first opposition stands for the value/disvalue quality that is represented along with the first order *Sinn*-fragment. Let us call these truth$_1$ and falsity$_1$, generically V_1. The second opposition concerns the internal harmony or disharmony of the full judgmental *Sinn*, i.e., of the *Sinn*-fragment and the value/disvalue quality represented along with it. Let us call these truth$_2$ and falsity$_2$: generically V_2. Then we have:

(i) a *Sinn*-fragment: <a, b, cause>

(ii) affirmation/negation: ascribing V_1 to <a, b, cause>

(iii) a judgmental *Sinn* <<a, b, cause>, V_1>

- a *Sinn*-fragment of the second order.

(iv) judgmental decision of the second order:

ascribing V_2 to <<a, b, cause>, V_1>

I would think, Lask over-hastily lumps together Bolzano's truths- and falsities-in-themselves and Husserl's sentence-meanings. Husserl's sentence-meanings are either (i) or (iii). Bolzano's truths- and falsities-in-themselves are the harmonious or disharmonious structures designated as truth$_2$ or falsity$_2$ respectively, abstracted from the subjective acts that constitute them.

We still do not understand what Lask means by saying that the pure logic of *Sinne* knows only oppositionally split *Sinne* and not the originary opposition-free *Sinne*. To understand this, we need to begin anew with one of Lask's central concerns in his theory of judgment.

IV

Beyond Oppositional Structure

Logic, according to Lask's interpretation of Kantian Copernican revolution, falls into two strata: the primary, objective logic, i.e., transcendental logic, and the secondary, non-objective logic. Judgment belongs to the latter. The task of a theory of judgment, in "the age of Kantianism", is to clarify the relation of judgment to the objective-level structure. This derived character of judgment is covered up by the widely held but erroneous view that, according to Kant, an object is constituted by judgments and that judgment should be assigned a primacy within the theoretical domain. According to Lask, judgment as such has no transcendental and objective significance. It is an entity of merely formal-logical significance. What is needed then is to go beyond the domain of judgment and to embed it within the larger objective, i.e., transcendental structure.

While judgment is removed from the objective structure, it is yet *somehow* directed towards the object. The objective element is somehow incorporated into the judgment. The judgment is at the same time both a distancing from the object and modifies or transforms the object into an "artificial" structure. A theory of judgment should build a bridge across this "distance" which separates judgment from object.

A. Layers of oppositionality:

Another feature of judgment, recognized since the beginnings of logic, is the oppositional character (*Gegensätzlichkeit*) of the judgmental *Sinn*. This is expressed by saying that a judgment is *either* true *or* false. For Lask, this feature is closely connected with the already mentioned "artificiality". What now Lask proposes to do is to contrast the oppositional structure of propositional *Sinn* (i.e., that it is either true or false) with an opposition-free original structure belonging to the transcendental-logical region. All logic of pure *Sinne* as well as all logic of *Geltung* knows only of such oppositional structures, and so has no inkling of opposition-free value. We still have to understand what Lask means by it.

It is commonly supposed that the oppositional structure (truth/falsity; validity/invalidity) is first introduced by the judgmental attitude which makes a decision for or against, about a value-indifferent presented content or matter. Lask maintains that even in the "matter" or substrate of judgmental decisions there is a valuational opposition.[20] To understand this thesis, let us take a few steps backwards and recollect some distinctions already made. First, there are the (psychological) acts of judging which may be correct or incorrect. Distinguished from these, are the contents or *Sinne* which are either $true_1$ or $false_1$. The $true_1$/$false_1$ meaning-structures, although detachable from the acts, first arise with those acts. The meaning-structures themselves, of whom '$true_1$' and '$false_1$' are ascribed and about whom judgmental decisions are made, are to be distinguished from the total object of judgmental decision which is the meaning structure + $value_1$ quality (= the total thought object). Correctness/incorrectness of the act needs a measure or standard; so does also the *Sinn*. Correct judging is correctly deciding about the *Sinn*-fragment. This *Sinn*-fragment together with the value-quality assigned to it constitutes the *Sinn* of the judgment. Thus we have three levels of oppositionality:

(i) correctness/incorrectness of the psychological acts.

(ii) $truth_1$/$falsity_1$ ascribed to the logical *Sinn*-fragment.

(iii) $truth_2$/$falsity_2$ of the structure against which both (i) and (ii) are measured. This structure must be independent of judging.

In judgment one decides *about* the $truth_1$ or $falsity_1$ of an in-itself $true_2$ or $false_2$ entity, so that $Truth_2$ and $Falsity_2$ obtain in-themselves. Truth-in-itself deserves affirmation and Falsity-in-itself deserves denial. The psychological act is incorrect when a false-in-itself is taken to be true-in-itself. But this requires that there be a truth and a falsity independent of both (i) and (ii).

What Lask is arguing for is that there must be an in-itself truth and an in-itself falsity. This suggests a suspicious hypostatization of the negative, of falsity. One wants to say that subjectivity must somehow be responsible for production of falsity. Lask rejects this straightforward position, and finds it evident that although error is subjective, the very possibility of error requires that what is worthy of denial must have its own in-itself being. That which is set aside in a denial as disvalue is not error, but must be an in-itself falsity worthy of being denied.[21] What Lask calls $true_2$ and $false_2$ structures are not the entire propositional *Sinn* — but only "matter" which underlies and somehow enters into the *Sinn*: they are the objective correlates of correct affirmation and denial. These are the "primary objects" of judgmental decision. And yet, these "primary objects" of judgmental decision, truth$_2$ and falsity$_2$, are different from objects *per se*, the latter being entirely beyond valuational oppositions. The following figure will help visualize Lask's complex theory:

Psychological		Logical		Primary objects of Judgmental Decision	Metaphysical
Acts		*Sinne*			Objects *per se*
/	\	/	\		
Correct	Incorrect	Truth$_2$	Falsity$_2$	T_2—F_2	or *Gegenstände*

T_2 and F_2 serve as measures for T_1 and F_1. T_2 and F_2 are measured against objects *per se*.

The primary object of judgmental decision is a complex structure whose elements must be characterized by in-itself-belonging-together or in-itself-not-belonging-together. What Lask has in mind is "material" agreement or disagreement of elements (and *not* formal compatibility or incompatibility). Judgmental decision does not first add a value or disvalue quality to an indifferent matter, it rather tries to make explicit the value quality which belongs to the in-itself structure. The philosophical idea that there is first a mere value-indifferent relation of being represented together amongst the elements of the whole (as, e.g., between "earth" and "moving") to which judgmental decision then adds a value quality exists only in the eye of subjectivity, not in itself. Connected with this idea of an indifferent, merely represented whole, is the idea of copula as an indifferent tie. According to Lask, such a relation, the mere copula as distinguished from belonging-together and not-belonging-together is a product of subjectivity which separates the quality from the total structure. The copula, on his account, is what remains over as "the nameless, uniform residue", when the value quality is removed in thought — the result of a "*Depotenzierung*" of the original harmonious or disharmonious connectedness of the elements within the object of judgmental decision.[22]

B. "Matter and Form": A hierarchy of concepts

This concept of "primary object of judgmental decision" leads to one of those theses for which Lask's theory of judgment became famous: a theory of many-layered form-matter distinctions, the so-called "*Stockwerk*" theory. Keep in mind the classical theory that in judgment, matter is brought under form (form, being the relation into which the material elements enter). Lask's theory retains this, but relativizes "matter". Relative to the judgmental *Sinn*, the "primary object" (with its component elements belonging- or not-belonging-together) is the "matter". Relative to the act of judgmental decision, the *Sinn* is the "matter". But, relative to the "primary object", the object *per se* or *Gegenstand* functions as "matter". What this shows is that Lask was as much concerned — as noted earlier — with emphasizing the "distance" of judgment from the domain of objects, as to show how nevertheless the objects "intrude into" or are "worked into" (*hineingearbeitet*) the judgmental domain, so that within the structure of judgment, the non-objective oppositions (correct/incorrect and thus $true_1$/$false_1$) come into contact with (and so are measured against) the objective opposition of $truth_2$ and $false_2$, i.e., of in-itself-belonging-together and in-itself-not-belonging-together, and, in the last resort, this last opposition refers back to the structure of the object *per se*. Naturally, this leads to the question, whether objects *per se* are opposition-free or not.

Just as the concept of "matter" is many-layered, so also that of "form".[23] In a broad sense, all that is logical is formal, but in this sense, to be logical (or formal) is to possess the character of pointing to something, or rather of being-in-need-of-completion (*Ergänzungs-und Erfüllungsbedürftigkeit*). Form in this wide sense points to matter. Matter remains under form. Within this very wide sense, one may distinguish between "form" in the sense of form of judgment and inference and "form" in the sense of categories. Lask calls the former "structural form" (*Strukturform*) and the latter "content form" (*Gehaltsform*). The category, according to Lask, does not simply relate some component elements. It is itself an element, the logical element, which is in one sense "matter" — the "logical matter" — but is also to be called "form" insofar as it is "pointed towards" the elements with which it enters into the object and so is "in need of supplementation."

C. Subject-predicate distinction

Lask rejects the merely psychological theory of subject and predicate (that the distinction between subject and predicate reflects the mere direction of enquiry) as well as a purely grammatical theory, and develops what he calls a meta-grammatical theory, which, following Aristotle, ascribes to the subject-predicate distinction a factual, real significance. If for Aristotle, the subject must be substance and the predicate an accident, for Lask the subject must be "matter" of knowledge (that which is given as *zugrundeliegende*) while the true predicate is the category. Knowledge, or judgment is the predication of categorical predicates of "matter". The two elements cannot exchange their places: hence the irreducibility of the subject-predicate distinction. Furthermore, since for Lask, as already noted, the "matter" - "form" distinction appears at many different levels, he holds that the general theory (namely, that the subject is the "matter" and the predicate is the category) holds good of all cognitions, including even philosophical cognition. (This is an improvement upon Kant, inasmuch as Kant's theory of knowledge did not extend to philosophical knowledge.)

Contrast now the grammatical theory with Lask's theory. In the judgment 'a is the cause of b', the grammatical theory takes 'a' to be the subject and 'being the cause of b' to be the predicate and the copula to connect the two. On Lask's theory, the subjects are the material elements 'a' and 'b', the predicate is the category 'cause', the copula relates the two: matter and form. In the case of 'a exists' or 'a is identical with itself', the subject is 'a', the predicate is the category (existence, or identity) and the copula connects the two.

Lask is not to be taken as saying that 'a' and 'b' in 'a is the cause of b' are pure, conceptually i.e. categorically unformed matter. Even if they are categorically formed, as they must be in the present case (for, 'a' and 'b' must stand for things or events), the category 'cause' — according to him — must be able to "penetrate" the categorically formed matter and reach the element of pure matter in 'a' and 'b'. Every knowledge, it is true, begins with a conceptual accomplishment, and seeks to further this process, but a categorization always refers back to the "pure matter" in the already conceptualized subject.

Let S be the subject in a judgment and K_2 the category that serves as predicate. But S itself, let us suppose, is pure matter (m) brought under category K_1. Thus:

$$[S \text{————->} \underline{K_2}] = [m(k_1) \text{————->} K_2]$$

Lask's point is that K_2 does not reach back to m.

Lask's theory is not that the categorial form is a form of the copulative synthesis, which Kant sometimes gives the impression of having held. For Lask, the category, although a form, is also one of the elements. The three elements in 'a is the cause of b' are 'a', 'b' and 'cause'. Furthermore, all categories are not relational, 'existence' for example, is not. The copula represents nothing other than "jene einformige Bezogenheit, . . . die sich als indifferente Unterlage des eigentümlichen, aber überall gleichen und einformigen Zusammengehörens und Nichtzusammengehörens der Strukturelemente erwiesen hat."[24] It stands for the belonging-together or not-belonging-together of the elements — in our example: of 'a', 'b', and 'cause' (adding a certain directionality, in order to distinguish 'a is the cause of b' from 'b is the cause of a').

Lask claims that his theory can give the best account of the so-called existential judgments. The really important points that he makes are two: first, existence not being a real predicate (as Kant held) and being a logical predicate, is — for Lask — a predicate *par excellence* (for the predicate should be a category, on his own theory). Secondly, it is very important for Lask that we distinguish between the sense of "being" which permits opposition (i.e., the objective correlate of correct affirmation) and existence as a category which is beyond oppositional structure and belongs to the object *per se*. He takes Brentano to task for precisely making this confusion when Brentano understands by "existence" the copulative, positive "being", that which is worthy of affirmation — nonexistence being the correlate of true negative judgment. Brentano overlooks, according to Lask, that being worthy of affirmation (*Bejahungswurdigkeit*) itself presupposes a "belonging together" of the category "existence" and the matter to which existence belongs. What is affirmed, then, is not existence, but the "holding good of existence". For Lask, existence is an opposition-free element. Positive or negative being of copula is added on to existence.

D. Beyond oppositional structure

Thus, on Lask's theory, every judgment consists in category-matter ('a' and 'b', in the above discussed example) and category ('cause'). To affirm its truth is to affirm their belonging- together or not-belonging-together. But such belonging-together or not-belonging-together can characterize only the region of 'objects of judgmental decision'. Can it be said to characterize the region of objects *per se*, beyond the judgmental domain? If not, then the objects *per se* must be taken to lie beyond the opposition between positive value and negative disvalue.

In 'a is the cause of b', what either does or does not belong together are 'a', 'b' and 'cause'. Note that causality itself is not a mode of belonging-together. There is no causal-belonging-together. Causality itself is neither belonging-together nor not-belonging-together.[25] It is opposition-free. The same is true of the "inherence" between a thing and its property. Inherence itself is opposition-free. In 'The pencil is yellow', the elements 'the pencil' and 'yellow' and the category 'inherence' belong together. Thus the structure of objects *per se* does not show oppositionality, whereas the structure of judgment does show it; and yet they have the same elements in an important sense.

But this last point should be correctly understood. First, to say that the object *per se* has the same elements is *not* to say that in the object *per se* the elements are separated. The elements must be separated from the category ('cause' in our example), and *then* brought together — thereby making both belonging-together and not-belonging-together (whichever is the case) possible. What else is responsible for this initial separation than the knowing subjectivity? In the second place, the "harmonious" structure (where the elements, matter and category, belong together) is as much removed from the object *per se*, as is the "disharmonious" structure (in which the elements do not belong together). The $true_2$ is as much "artificial" as is $false_2$.[26]

In the object *per se*, there is no belonging-together of elements, but rather a simple being-in-relations. Belonging-together, in Lask's theory, is oppositional, i.e., has the opposite possibility of not-belonging-together: both characterize — at one stroke, as it were — the region of judgment, as possibilities. But the being-of-the-material-in-the-category, which is the object *per se*, is value-neutral. In Lask's words, what we have is "das schlichte, durch keinerlei Zerreissung hindurchgegangene Stehen der dortigen Elemente."[27]

Third, to speak of a *relation between* category and category-matter is misleading. A category, being a form, intrinsically points to something other than itself. Matter is so-called insofar as it is subject to a form. So quâ matter and quâ form, the two must already be together. To speak of them as *relata* is to treat the category as pre-formal and the matter as pre-material something.

With this last point, Lask takes another step towards the idea of an opposition-free being, against which the judgment, with its oppositional structure is measured. The step consists in formulating what he calls the *Principle of Categorial Differentiation*.[28]

Form, quâ form, points to matter. Matter, quâ matter, comes under, is subject to, form. Consider Form in general. As Form in general, it points to matter-in-general. It does not point to any specific, determinate sort of matter. A specific form such as 'causality' would, however, contain a pointing towards a specific sort of matter, such elements as can have causal relation amongst them, not, e.g., abstract numbers. A specific form has a specific character or content which Lask calls "*Formgehalt*", also its "*Bedeutngsbestimmtheit*", meaning-determination, that which is the meaning of 'cause', for example. Since a specific form also contains a pointing towards a determinate sort of matter, that specificity of matter by being directed towards which Form in general becomes a specific form is called "*bedeutungsbestimmendes Moment*", i.e., the meaning-determining aspect. Thus corresponding to the specific form "reality", there is the meaning-determining aspect (of matter) "sensuous-intuitiveness in general". Corresponding to the specific form "cause", there is the meaning-determining aspect "temporal before and after". Thus there is the following correlation:

Form-in-general <———————————> Matter in general

Form$_1$ <————————> M$_1$ ——————> (a_1, b_1, . . . X_1)

Form$_2$ <————————> M$_2$ ——————> (a_2, b_2, . . . X_2)

Each of these meaning-determining aspects comprehends all those infinite particularities which possess it, so that within each specific form, relatedness not merely to specific aspects of matter but also to a whole class of particulars (those which share that aspect) is already pre-delineated. Similarly, from the other side: take any material particular, by its very nature its being-subject-to the appropriate forms is already pre-decided.

Now consider the original Form, i.e., Form-in-general. It is supersensible and one. It points to something-in-general. In this situation, there being a seamless unity — there is no scope for harmony or disharmony amongst elements, so no oppositional structure. This is the opposition-free *Urregion*. Each specific category may be represented as a ray stretching from a specific form to a specific meaning-determining aspect in matter. But this directedness is already contained in that category's meaning-content. Since there is no separation between the two, there is no question of harmony or disharmony between form and matter. There is only: from one side, a pointing towards, and from the other side, a being-subject-to.

Whereas in the original object *per se*, the category and category-matter are in a seamless unity, in a judgment they "become" elements tied together as subject and predicate.

We had earlier noted that according to Lask, the pure logic of *Sinne* (of Bolzano and Husserl alike) knows only oppositionally split *Sinne*, but not the originary, opposition-free *Sinne*. Likewise, Lask thinks that the so-called *Geltungs*-philosophies of the Neo-Kantians know only of the oppositional structure value/disvalue, but not of the originary, opposition-free value. Now we can have some understanding of what he means.

The original object *per se* is opposition-free. Can we call it *Sinn*?[29] If it is *Sinn*, then it is the original, opposition-free *Sinn* from which judgmental *Sinn* is derived. Lask thinks there is nothing against such a usage. It seems to me that there is. If the *current* terminology is adopted, it may be called the ultimate *referent*.

What is the original, opposition-free Value? It must be located within the structure of the object *per se*. The elements in the object *per se* — even if not artificially separated — are the sensuous-intuitive ones. Value or Norm can lie only in the supersensuous Form in general which, as related to those specific elements, 'becomes' a specific form-content or category. Whereas material truth/falsity concern the elements, the Value concerns the Form in general. One would still want to know, why Lask considers the supersensible Form in general to be the original opposition-free Value.

V

Rickert and Lask

In order to correctly understand Lask's rather complicated theory, it may be of help to take a look at his relationship to Heinrich Rickert's theory on the same issues. I will not elaborate upon these theses of the school of Neo-Kantianism to which both belonged. The general distinction between *Sein* and *Sollen*, the interpretation of Kantian philosophy as a philosophy of *Geltung*, the bringing together of Kantian theory of category and logical theory of judgment in a manner that is closer than that implied in Kant's *Metaphysical Deduction* — all this Lask owed to Rickert amongst others. Both Lask and Rickert found Husserl's notion of *Sinn* useful, and Lask, of all Neo-Kantians, appropriated it into his own thinking. However beyond this, Lask went his own way in a direction that Rickert found unacceptable.

I think the main issue between them came to be: is the fundamental idea of value essentially oppositional, i.e., does 'value' imply 'disvalue'? In general, are the oppositional structures truth/falsity, *Sinn/Sinnlos* the final truth about those concepts, or is it not only possible but necessary to recognize a Value which is beyond the opposition of value-disvalue, a truth that is beyond the opposition of truth-falsity, a *Sinn* that is beyond the opposition of *Sinn-Sinnlos*

— also a Being that is beyond the opposition of positive being (affirmation-worthy) and negative non-being (denial-worthy). Lask strongly affirms such a transcendence of all oppositional structures, Rickert denies it. The originality of Lask's philosophy lies in the general *Stockwerk* theory of matter and form as well as in this idea of transcendence.

Rickert's insistence that valuational oppositionality is ultimate is traced by Lask to a deep misconception of the nature of Kantian transcendental logic, and consequently to a misplaced emphasis on judgment in the overall structure of transcendental philosophy. If Kant extended the domain of logic beyond formal logic, that only means that logic, i.e., transcendental logic, is the logic of objects *per se* and the forms of transcendental logic, i.e., the categories constitute object *per se*. While this much is acknowledged, it is subjected to two misinterpretations. In the first place, because of Kant's argument in the metaphysical deduction, the forms of judgment themselves are assigned this transcendental-constitutive role, whereby judgment itself has come to acquire a central place in transcendental logic. Lask's work is directed against this misinterpretation — as has been noted earlier in this paper. Furthermore, since Kant regarded transcendental logic to be concerned with the possibility of objective validity, it is understandable that objective validity is equated with truth. Consequently, logic is taken to be concerned with the presuppositions of true but not of false judgments. Lask also rejects this interpretation of Kant.

I would like to add a few remarks with regard to this second misinterpretation of Kant. Quite irrespective of the validity of Lask's own construal of the Kantian transcendental philosophy, it seems to me to be undeniable that 'objective validity' should not be identified with 'truth'. For, false judgments are also 'objectively valid', they too involve categorial synthesis. It is preferable to say that objective validity assures either truth or falsity, in either case objectivity and so conformity to the rules of understanding. So far, Lask's criticism of Rickert is correct.[30] But where does one go from here? One alternative is to suspend the question of truth and falsity, and to take only the *Sinne*, the ideal-objective meanings, of the judgments (true or false), and make logic a theory of pure meanings. This is the alternative Husserl chose.[31] But then one is either giving up the key concept of *Geltung* or *Value* or making it almost ineffective if anything of it at all remains. Lask chooses another direction. Behind and beyond the opposition between true and false, he posits as the last measure an opposition-free being, and calls it also *Sinn, Wert* as well as *Gegenstand*. It is to this move that Rickert objects.

Rickert's objection is not merely about the correct application of the words "Sinn" and "Wert", but is also substantial: Lask wrongly separates the idea of value from that of *Sollen* on the ground that the latter contains a reference to the evaluating subject and is clearly oppositional. Rickert insists that the two concepts, 'Value' and '*Sollen*' are so closely connected that one cannot deny the oppositionality of the former.[32] There is a sense in which Rickert will recognize that the opposition between value and disvalue is contained within value, i.e., there is a sense of 'value' in which both value and disvalue come under it. But this is a sense of 'indifference' rather than genuine 'transcendence'. Lask is after genuine transcendence that is (not indifferent to and abstracted from differentiation but) *prior to* differentiation.[33]

For anyone thinking from within the tradition of transcendental philosophy, it is unavoidable to ask, how much of transcendence is compatible with that tradition to begin with. Lask remarks that the only transcendence Rickert knows of is: "being absolutely worthy of affirmation" and independence of subjectivity in this sense, also detachability of *Sinn* from its real substratum.[34] But neither he nor Husserl know of genuine transcendence of the object *per se* or *Gegenstand*.[35] To characterize Value as being-worthy-of-recognition (*Anerkennungswürdigkeit*) is already to objectify it and thereby to insert it into the subject-object relation.[36] To be an object is to be a *Sinn* "floating before" subjectivity.[37] Even "opposition-free" *Sinn* is at one place taken by Lask to be an "immanent" something — only it is "next to" the genuinely transcendent *Gegenstand*.[38]

In the long run, the issue becomes: are affirmation, and its objective correlative, positive being, adequate to characterize transcendent being? Lask puts affirmation and negation on the same plane, true judgment and false judgment on the same level of distance, the same level of "artificiality", and wants to characterize the transcendent *Gegenstand* as beyond both, i.e., beyond the reach of judgmental knowledge. Rickert recognizes that there is a sense in which truth and falsity are correlates and that just as there are true *Sinne*, so also there are false *Sinne* — equally objective, equally detachable from the subjective acts. But even then he would assign to the affirming, positive, consciousness a priority for the conception of *theoretical* subjectivity as for the idea of *Gegenstand*, i.e., object *per se*. To clarify this, Rickert distinguishes between two levels of affirmation: one which, like negation, is a judgmental decision in response to a prior question, and the other prior to and independent of questioning ("*das frageloses Ja*").[39] Lask recognizes only the former, and rightly finds in it an opposition to the negative judgmental decision, but he does not see that the latter is indispensable to all theoretical subjectivity, and must be the correlate of an opposition-free reality.[40]

VI

Some Questions

Some questions, briefly hinted at earlier in this essay, may now be formulated.

First, one suspects that Lask's transcendent *Gegenstand* would have to be construed as an unknown and unknowable thing in itself, unless Lask tells us how it is known. If it is known, that knowledge cannot be judgmental. Lask characterizes it as "*übergegensätzliches, überurteilsartiges Erkennen*", a supra-oppositional, supra-judgmental knowing, as merely receptive to the *Gegenstand*.[41] The transcendent *Gegenstand* has to serve on Lask's theory as the ultimate *measure* of the truth or falsity of our judgments. This may mean either of two things: *either* that in all our decisions about truth or falsity of our judgments we *do* in fact *measure* against the original *or* that even if we do measure against other criteria (such as coherence of judgments or pragmatic success), the ultimate *reference* is to the transcendent *Gegenstand*. If it is the latter, then Lask's argument that unless the transcendent *Gegenstand* is supra-oppositional we cannot avoid an infinite regress, would not be compelling. That argument is compelling only if we choose the first alternative, but then we are faced with two interpretations of that alternative. By saying that we do in fact measure against the original, Lask means that this evaluation does preserve, inasmuch as it is intended to account for, the truth or falsity ascribed to our pre-philosophical judgmental decisions (to the effect, e.g., that a is the cause of b), *or* that this evaluation, metaphysical as it is, casts doubt on the ultimate validity of any judgmental decision. If, of these two interpretations, we choose the second, then also the transcendent *Gegenstand* does not really serve as a measure for our decision, if a is the cause of b or not. Only on the first interpretation, it does, but then one would want to know what is that passive reception, that suprajudgmental knowing of the being-together (of the elements and the category, not put together by thought after being separated, but the original, as yet undifferentiated being-together) which we have at our disposal such that we use it to measure the truth or falsity of our judgmental decision as to whether a is the cause of b or not. Ordinary, non-judgmental perception (when we perceive a thing) provides the only available instance, but for obvious reasons there is no conceivable means of perceiving, the original being-together of category-matter and category in this case. It is possible that Lask had in mind some other, metaphysical sort of knowledge of the transcendent being, but then it is difficult to see how that knowledge is actually the measure for our judgmental decisions in a manner that *preserves* their truth values. One may suggest, what is the measure, on Lask's theory, is not that knowledge, but the being itself, but then being is better construed as the ultimate *referent*. Its function in relation to judgment would then be different.

Secondly, one wants to know if the transcendent *Gegenstand* which is the original *Urbild* of the judgment 'a is the cause of b' is the same as, or different from the transcendent *Gegenstand* which is the original *Urbild* of the judgment 'a has the property Φ'. What, in other words, one may be asking is, are there many such transcendent *Gegenstände* or only one? Lask, from all accounts, tends towards the monistic alternative — which gives his theory a better metaphysical status, but makes it epistemologically less useful. The monistic tendency — almost Neo-Platonic — is nowhere clearer than in the thesis that the original supra-oppositional Value is the Form-in-general.

Finally, I want to express my anxiety about to what extent, if at all, and in what sense, one is justified to speak of a transcendental-*logical* constitution of the transcendent *Gegenstand*. The mere being-together of category form and category matter, prior to any differentiation and so prior to any categorial synthesis, does not appear to be a proper subject matter even of a transcendental *logic*. What, in effect, Lask may be suspected of doing is to ground transcendental logic on a metaphysics of Forms, and yet continue to call this latter Neo-Platonistic thesis of differentiation of Form-in-general into a plurality of specific forms a transcendental logical thesis. These criticisms do not affect the fascinating and important *Stockwerk* theory, the many-layered theory of form and matter, the many-tiered theory of true-false oppositionality, and the correspondingly structured theory of *Sinne*.

NOTES

[1] Henry Allison, Kant's Tanscendental Idealism, pp. 68f.

[2] I. Kant, *Critique of Pure Reason*, A78 = B104.

[3] Ibid, A55-56 = B79-80.

[4] H. Wagner, "Zu Kants Auffassung bezüglich des Verhaltnisses zwischen Formal und Transzendentallogik", *Kant-Studien*, 68, 1977, pp. 71-76.

[5] H. Rickert, *Der Gegenstand der Erkenntnis*, 6th edition, Tübingen: J.C.B. Mohr, 1928, p. 271fn.

[6] H. Rickert, "Zwei Wege der Erkenntnistheorie. Transzendentalpsychologie und Transzendentallogik", *Kant-Studien* 14, 1909, 169-228.

[7] E. Lask, *Die Lehre vom Urteil*, Tübingen: J.C.B. Mohr, 1912, p. 156.

[8] Ibid., pp. 4-5.

[9] Ibid., p. 90.

[10] I bid., pp. 148-9.

[11] Ibid., p. 117.

[12] Ibid., p. 5.

[13] Ibid., p. 7.

[14] Ibid., p. 161.

[15] Ibid., p. 167.

[16] Ibid., pp. 168-9.

[17] Ibid., p. 193.

[18] Ibid., pp. 170-1.

[19] Ibid., p. 172.

[20] Ibid., pp. 17-18.

[21] Ibid., pp. 21-22.

[22] Ibid., p. 37.

[23] Ibid., pp. 55-56.

[24] Ibid., p. 75.

[25] Ibid., pp. 91-3.

[26] Ibid., p. 95.

[27] Ibid., p. 98.

[28] Ibid., pp. 120f.

[29] Ibid., pp. 132-33.

[30] Ibid., pp. 155f.

[31] E. Husserl, *Prolegomena to a pure logic (Logical Investigation)*, Vol. I, E. tr. J.N. Findlay, Humanities Press.

[32] H. Rickert, *Der Gegenstand der Erkenntnis*, p. 337.

[33] *Die Lehre vom Urteil*, p. 143.

[34] *Die Lehre vom Urteil*, p. 156.

[35] *Die Lehre vom Urteil*, p. 172.

[36] E. Lask, *Gesammelte Schriften*, III, p. 92.

[37] E. Lask, *Gesammelte Schriften*, III, p. 85.

[38] E. Lask, *Gesammelte Schriften*, III, p. 103 fn. 2.

[39] Rickert, loc. cit., p. 341.

[40] Ibid., p. 342fn.

[41] Lask, *Die Lehre vom Urteil*, pp. 135-6.

9. HUSSERL ON "POSSIBILITY"*

My investigation into the nature of the modal concepts Husserl employs is motivated by the expectation that we may thereby be able to gain a deeper understanding of some of his crucial philosophical moves. The immediate stimulus for such an enquiry, however, was provided by Jaakko Hintikka's book *The Intentions of Intentionality* in which he attempts an interpretation of Husserl's notion of intentionality from the standpoint of possible world semantics.[1]

As the title of this essay indicates, my primary concern shall be the modality of possibility, and only secondarily the modes of actuality and necessity. As is well known, necessity may be defined in terms of negation and possibility: p is necessary if the negation of p is not possible.

1. SOME TEXTS

Let me begin by quoting a series of statements from Husserl, which would help us to appreciate not only the centrality of the concept of possibility in his thinking, but also the great variety of senses in which he uses this modal concept.

1. "... possibilities themselves are ideal objects. Possibilities can as little be found in the real world, as can numbers in general, or triangles in general."[2]
2. "Individual Being of every kind is, to speak generally, 'accidental'. It is so-and-so, but essentially it could be other than it is." (Hua III/1, 12)
3. "To each essence there corresponds a series of possible individuals as its factual instancings." (Hua III/1, 20)
4. "... 'the real world', as it is called, the correlate of our factual experience, then presents itself as a special case of various possible worlds and non-worlds, which, on their side, are no other than correlates of the essentially possible variations of the idea 'empirical consciousness'." (Hua III/1, 100)

---

* Originally in: *Husserl Studies* 1: 13-29 (1984).

5. "All actual experience refers beyond itself to possible experiences, which themselves again point to new possible experiences, and so *in infinitum*." (Hua III/1, 02)
6. "... every actuality involves its potentialities, which are not empty possibilities, but rather possibilities intentionally predelineated ..."[3]
7. "Without such 'possibilities' 'I can always do so again' there would be for us no *fixed and abiding* being, no real and ideal world."[4]
8. "Nothing speaks for (the proposition) that the world is not, and everything for (the proposition) that the world is. But, what is important is that this perfect empirical certainty, this empirical indubitability, as empirical, still leaves open the possibility that the world nevertheless is not ... " (Hua VIII, 54).
9. Transcendental subjectivity is the "universe of possible meanings".[5]

In these texts, we have heard Husserl tell us about pure possibilities; possibility of an individual fact's being otherwise; the possible instancings of an essence; of possible worlds; of possible experiences; of potentialities; of the fundamental possibility indicated by the locution "I can always do so again"; of the possibility of the world's utter dissolution into nothing; and, finally, of the transcendental subjectivity as "the universe of possible meanings". What I will attempt, in the first place, is to bring an order -- systematic as well as historical -- into this great variety of locutions.

2. "PURE POSSIBILITIES"

Pure possibilities, at their purest, i.e. as independent of any reference to real spatio-temporal being, pertain to species or essences. Only derivatively, would Husserl say that for every essence, there is a (pure) possibility of individual instances. The concept of pure possibility, as formulated in the *Logical Investigations*, comes closest to Leibniz's. Aron Gurwitsch interprets Leibniz's concept as follows:

> Pure possibilities, according to Leibniz, are the objects of incomplete concepts, of concepts which figure in the rational and abstract sciences. Such concepts determine the objects to which they relate only in certain, but not in all respects. The object corresponding to a complete concept is fully determined by its concept, i.e. is fully individuated. The contrast is between 'Triangle' and 'The triangle drawn on the blackboard here before me', or between 'vague Adam' and 'possible Adam' (as a possible variant of Adam). The objects of such incomplete concepts are not capable of existing by

> themselves. The objects of complete concepts are capable of existing. The former are therefore pure possibilities. The adjective 'pure' serves a double purpose: it signifies not only independence of spatio-temporal positing, but also the fact that these objects are simply not capable of real existence. However, for Leibniz as well as for Husserl, these pure possibilities possess a sort of actuality that is specific to all abstract objects of their kind. In Husserl's terminology, they are ideal objects. As Leibniz insists, they are 'discovered'; the truths grounded in them are eternal and necessary truths. Leibniz proceeds, nevertheless, to ground their specific sort of actuality in the being of another actuality i.e. in God's intellect. Husserl does not, at least not immediately, make this move.[6]

The ideal objects that are pure possibilities, then, are what Husserl calls "dependent essences" or "*abstracta*" (Hua III/1, 35). Thus, "red circle" refers to a dependent essence, it cannot be by itself. Any individual red circle has more determinations than merely being red and being a circle. That a pure possibility according to Husserl is an *abstractum* or a dependent essence is borne out by the facts that necessary truths on his view are about pure possibilities, and that, in the third Logical Investigation, he defines synthetic necessary truths in terms of dependent objects.

If 'A' and 'B' designate two dependent essences, to say that 'AB' designates a pure possibility is none other than saying that there *is* a compound essence AB. There is such an essence and hence such a pure possibility, if its component essences, A and B, are consistent. If A is a simple essence, then it is not consistent with itself, and therefore a pure possibility. Husserl rightly insists that whether two essences are compatible or not, depends upon what whole, or what sort of unification, is under consideration. A and B, then, may be compatible in relation to a whole W (and not compatible in relation to a whole W^1). The same holds good of incompatibility. What then is a pure possibility is not mere AB, but W(AB).

Further, the idea of compatibility that is being used for our present purpose, needs to be *independent* of the formal-logical notion of "freedom from contradiction". For the logical forms themselves are abstract essences, and compound logical forms, if consistent, are pure possibilities; and we are seeking for a notion of pure possibility, of which, "logical possibility" is a species. At this stage, therefore, Husserl gives an account of "compatibility" in terms of "fulfilling sense". A compound meaning is consistent if the act intending it has a fulfilling sense, i.e. if the partial intuitions of the component meanings can be combined into a total intuition. When no such unified intuition is possible, the components are incompatible. When the meaning is consistent, the essence/species referred to has "ideal existence", i.e. *is* a pure possibility.[7]

There are two features in this thesis to which I wish to draw attention at this point. In the first place, in the domain of essences, possibility, i.e. pure possibility, amounts to (ideal) *actuality*. To be possible is to be actual. The two modes coincide.[8] It appears that such a coincidence does not hold good of the real world of individuals and facts, where, although the actual is also possible, all that is possible is not actual. About this intermodal relationship in the domain of real being, I will speak later. For the present, the coincidence of modes in the domain of essences is significant. The very being of essences is the being of a possibility. I have recalled earlier that Leibniz grounded this "actuality" of pure possibilities in the intellect of God. We shall ask later on, if Husserl needs such a recourse.

The other thing to be noted is that the explanation of (pure) possibility in terms of intuitability appears to be tantamount to reducing the objectivity of the former to what is a subjective notion. It would seem that what can or cannot be intuited depends as much on the object under consideration as on the subject who is trying to intuit. If it be said that "intuitability" for Husserl is an objective notion, then we are back with a new pure (objective) possibility, "W(AB) with respect to a unity of intuition I", and the being of this pure possibility, i.e. of I(W(AB)) would need explication in terms of *its* intuitability. There is thus an incipient infinite regress on our hands.

3. "REAL POSSIBILITIES"

(a) An independent essence or *concretum*, which completely determines its object, is the essence of an individual. As Gurwitsch tells us, Leibniz considered complete concepts alone as having objects that are capable of real existence. In Husserl's words, an independent essence or *concretum* is the essence of an individual that is capable of real existence (Hua III/1, 35). Thus, whereas an abstractum is only a pure possibility, a concretum *has real* possibility. The individual whose essence it is may as well have been a really actual individual. Whereas this essence, the concretum, has ideal actuality, the individual it determines is a possible individual that may or may not be a real existent. What is possible in this sense (of real possibility) then, is not the essence, the concretum, but the individual it determines. A variation in phantasy of a real individual (or of my own ego) would result in an individual which could have been real. Socrates six feet tall and without a snub nose but with a straight one, would be such an individual with real possibility; he is capable of existing. This is what Husserl often calls phantasy-possibility. When therefore he says, as in the second of the quoted texts, "Individual Being of every kind ... could be other than it is", he means: Take any individual reality, imagine variations in its concretum, i.e. in its essential properties -- properties which make him *this* individual -- and you come up with another individual which very well might have been real.

There is a slight ambiguity here, generating a problem of great concern for many contemporary philosophers, which Husserl does not appear to have taken note of. There are two senses in which one says "A might have been other than what it is". In one sense, "A might possibly have had some property or properties different from what it in fact has, and would still have remained A". In another sense, the possibility of A's being otherwise would have been so radical that we would not any longer say of this possible individual that it is A, we would rather say it is some other individual. But, in both cases, we are dealing with possible individuals -- possible in this first, and broadest sense of "real possibility". The problem then is: what imagined variations of A are those which are compatible with our saying that they are still variations of A? How radical the variations have to be, so that we would rather say it is not A any longer? Thus the possibility of A being other than what it is, when radicalized, is the possibility of A's ceasing to be A, i.e. the possibility of A's ceasing to *be*. These are all *real* possibilities in the broadest sense, as contrasted with the 'pure possibility' which is none other than the ideal existence of abstract or concrete essences themselves. I will call this 'real_1 possibility'.

(b) The third text quoted presents a somewhat different problem. "To each essence there correspond a series of possible individuals as its factual instancing ... " Consider an abstractum such as "philosopher". This dependent essence cannot be by itself, the philosopher has to be a human, then a Greek, an Indian or an American, and so on and so forth, till we have one of many possible *concreta* or independent essences. Each such independent essence would determine an individual, which is a possible individual: it can be, or could have been, real. To each abstract essence, then, there correspond many possible individuals inasmuch as each abstract essence is a component of many different concreta. The nexus of *abstracta* constituting a concretum is not necessarily so; here, as noted earlier, ideal existence is nothing but compatibility with respect to a certain whole.

What about a concretum which is an independent essence but also an eidetic singularity? Should we say that to each concretum there corresponds a series of *possible* individuals as its factual instancings? There are, so it seems to me, two ways of understanding the notion of concretum, such that under one formulation, the question has to be answered in the negative, under another in the affirmative. An affirmative answer requires that the same concretum could be the essence of more than one individual, that, e.g., Socrates and a variant of Socrates (with a minimal variation, of course) have the same essence. Husserl does not seem to have had this in mind. To my mind, Husserl so understood "concretum" that it is the essence of one and only one individual. If this is the case, then, the thesis that an essence may have many possible instances does not hold good of an independent eideitic singularity. It is true of abstract essences alone.

(c) I have called the possibility of an individual as determined by a concretum "$real_1$ possibility", but this does not appear to be the standard usage amongst philosophers including Husserl. I called it "$real_1$ possibility", only to contrast it with the pure possibilities which, being dependent essences, do not even determine an object as completely as is needed for that object to be able to have real existence. When Leibniz held that the world of possibles as the sum of all those things that are capable of existing is wider than those that really do exist, he had this concept of possibility in mind. However, according to the standard, i.e. more common philosophical usage, "real possibility" carves out, from within the Leibnizian domain of possibilities, a subset which however is still larger than the real, actual world. Only the principle by which this subset, that is a proper subset of the Leibnizian domain and yet includes within it the real world as a proper subset, is carved out is different in different philosophers.

Leibniz already has a principle of restraint on the wide domain of possible individuals: every individual must belong to a definite world, i.e. a domain of individual substances whose unity rests on a fundamental concept specific to that world. Each such world is a compossible system, only one such is the real world. The concept of real possibility, then, that is more restricted than the one expounded earlier is the concept of compossibility of really possible individuals in one world.

It must be emphasized that Husserl never explicitly rejected the Leibnizian thesis. On the contrary, there is every indication that he never ceased to be fascinated by it. In *Ideen* III, we find the following statement: "It is actually the case, as the old rationalists held, that an infinite plurality of possibilities precede actuality" (Hua V, 56). In §60 of the *Cartesian Meditations*, an attempt is made to give sense to the Leibnizian thesis. "Naturally Leibniz is right", Husserl writes, "when he says that infinitely many monads and groups of monads are conceivable but that it does not follow that all these possibilities are compossible; and, again, when he says that infinitely many worlds might have been 'created', but not two or more at once, since they are incompossible."[9] I will return to his way of making sense of Leibniz in a later context. For the present, still maintaining the ontological naiveté which has characterized this discussion until now, we may say that Husserl did believe in the Leibnizian thesis about many compossible worlds of which the actual is one. What we still have to find out is, what he meant by "compossibility"?

The infinitely large and chaotic domain of $really_1$ possible individuals has now been ordered into infinitely many unitary systems of them, the many compossible worlds, each of which is *really* possible in a more restricted and proper sense. For this latter sense, I will use '$real_2$' and its derivatives. What now is the relation of the one actual world and these many *really*$_2$ possible worlds? The well-known Leibnizian strategy is not available to Husserl. A phenomenological philosophy cannot tell a story of creation, it cannot have any account of *how* and *why* the only actual world, the one which is ours, could, from amongst a plethora of competitors, come to acquire the status of actuality. It will, however, introduce a way of dealing with this question according to which one asks about the origin of *sense* rather than the origin of things, and in which the philosopher ceases to speak merely "blindly" of "creation" and such things, and instead clarifies meanings with intuitive insight. Again, I will postpone considering this till later in this essay.

(d) A further restriction on real possibility is now in order. Given our real world as it is structured by its laws and general *types* of things and events belonging to it, one may speak of various possibilities with regard to those things and events, consistent with those typifications and laws. In other words, we are thinking of possible variants of this real world such that the general typifications and/or laws obtaining in this world are preserved in those variants. These variants do not exhaust all compossible worlds, but -- needless to say -- each such variant is a compossible world. It may be said, this real world, the one which is ours, contains these possibilities inasmuch as these possibilities are not just any of the real possibilities, but only those which are consistent with the laws of this world. Let us call these physical possibilities without claiming that this world of ours is all physical in nature. Saying that the real world *contains* such or in fact any sort of possibilities has been denied by Nicolai Hartmann. Drawing upon a Megarian concept of possibility, Hartmann has given a powerful defence of the position that just as in the ideal world of essences possibility and actuality coincide, so also in the real world, what is possible is actual and what is not actual is not possible.[10] I have elsewhere examined this position and found it unacceptable and so will not stop to consider it now.[11] This certainly is not Husserl's position. For Husserl, the real actual world is shot through with possibilities of various kinds. He will not be averse to the notion of 'physical possibility' that I have introduced. On the contrary, much that he says seems to imply such a notion.

4. HUSSERL'S PHENOMENOLOGICAL APPROACH

There are, in the history of Western thought, three broad conceptions of possibility, each allowing for a wide range of variation in formulation. These are: The Aristotelian concept of potentiality, the Leibnizian concept of pure possibility, and the Kantian critical theory of the modalities. The first makes possibility qua potentiality a stage in the development of the actual. The second makes the actual a selection out of the many possibles. For the Kantian critical theory, the three modalities -- possibility, actuality and necessity -- do not pertain to different stages in the development of an object, nor do they pertain to less and less inclusive regions of being (so that more is possible than is actual and more is actual than necessary). They in fact are not determinations of the object itself, they concern only *how* the object, together with all its determinations, is related to the understanding, to empirical judgment and to reason.[12]

Although the picture outlined in this paper up to this point makes Husserl seem somewhat of a Leibnizian, that impression needs to be supplemented by bringing out the Aristotelian and the Kantian elements in his thoughts about the modalities. For the Aristotelian strain, consider the sixth text quoted by me: "every actuality involves its potentialities, which are not empty possibilities, but rather possibilities intentionally pre-delineated ... ". The Kantian strain is nowhere more clearly to be discerned than in the distinction between *Sinn* and

Satz in §133 of the *Ideen* I, wherein the modalities are said to concern the modes of positing and not the objective *Sinn*. However, each of these components of Husserl's thoughts on possibility, and on modalities in general, undergoes transformations in accordance with his phenomenological method of philosophising. And to this transformation, I shall now turn.

If a spatial metaphor be permitted, then I should say that the account given in this section is *from above*. We started with pure possibilities, and then descended through narrowing spheres, and corresponding concepts, of possibility to the real actual world which is ours, and then to the possibilities belonging to it, or *contained in it*. This scheme is ontological.

Phenomenology by the very nature of its enterprise *cannot* follow this route. If its concern is with *senses* and not with *things*, and if senses are constituted in, and need to be clarified with reference to, consciousness, then phenomenology has to begin, taking up the same spatial metaphor, *from below*. But we shall soon find that the words "above" and "below" are misleading, not for the obvious reason that they are spatial metaphors but for more important philosophical reasons.

I have to begin with my consciousness, and clarify the 'origin' of the basic modal concepts and various patterns of intermodal relationships. The distinction between the ontological and the phenomenological approaches is thus stated by Husserl in a text, important for my present purpose, thus:

> " ... dass der Ontologe die Idee der Nature als eine Art ideale Wirklichkeit nimmt, dass er sich auf den Boden dieser Idee stellt, um sie nach ihren Eigentümlichkeiten zu erforschen; ... während der Transzendentalphänomenologie nicht in einem seinden idealen Raum sein Thema hat, sondern in der Idee eines Bewusstseins überhaupt .... Er fällt kein geometrisches Urteil, sondern Urteile über all die transzendentalen Möglichkeiten, auf denen selbst die Möglichkeit geometrischer Urteile und geometrisch wahrer Urteile beruht." (Hua XI, 222)

It is at this point that I need to dispose of an objection that is likely to arise. If, as a transcendental phenomenologist, Husserl's interest is in how the senses "possible" and "actual" originate in the domain of transcendental subjectivity, then why was the Leibnizian account so far given at all necessary? Is it the case that the idea of pure possibility, and of the many compossible worlds, and all that, is a vestige of that essentialism which Husserl, as he became more and more of a transcendental phenomenologist, gradually gave up? In the later writings, he talked more of "open possibilities" and "motivated possibilities" than of "pure possibilities" as ideal objects. Now, how essentialism is related to transcendental phenomenology is a larger issue on which I will not comment in this paper. But it will suffice for my present purpose to insist on the following:

It is a well-known thesis of transcendental phenomenology that constitution analysis shall begin with the constituted sense as its guiding principle. The ontological account of possibility was meant as providing us with a picture of the *senses* of "possibility" and "actuality" *as they stand fully constituted.* Ontology represents the constituted achievements considered apart from their constitutive origin. Phenomenology has to overcome this naiveté. To overcome a naiveté is not to reject an error, it is to comprehend that naiveté quâ naiveté. Transcendental phenomenological account of the modalities has to do this, but in doing so it cannot help identifying the constituted meanings themselves.

An account of "possibility" in accordance with the plan of constitutive phenomenology has to recognise the following main stages:

(a) In simple perception, a thing, the thing perceived, is bodily there before me. It is there before me with a certain primitive certainty. We do not say, it is "really" there, or "actually" there. It is simply there. The objective sense (*gegenständlicher Sinn*) component of the noema and the modality of being (*Seinsmodus*) are not yet distinguished. This is the original, unmodified certainty.[13]

(b) This original certainty may be interrupted by some disharmonious course of experience. What was given originally as a human being may subsequently be given as a shadow or then again as a human figure. In such a case, the thing begins to appear questionable, doubtful, disputable. The two conflicting characterisations, "a human being' and "a shadow", come to have the same mode of being "questionable". The intentions implicit in the original perceptual apprehension are obstructed. I vascillate between the apprehensions: "a human being" and "a shadow". Thus already at the level of pre-predicative experience, we have a modalisation of the original belief: the two alternatives before me are "problematical possibilities".[14] In an important sense, this is the origin, in pre-predicative experience, of the concept of possibility. As Husserl writes: "Wo immer ein Bewusstsein den Modus Gewissheit verloren had und in Ungewissheit übergegangen ist, ist auch die Rede von Möglichkeiten" (Hua XI, 39). If in the original perception, "Objective *Sinn*" and "mode of being" are not separated, it is in the transition to conflict and disharmony, that they become separated. In fact, by the title "mere objective *Sinn*" something is designated, which was not there beforehand in simple perception (Hua XI, 229). The consciousness of this distinction is a new experience.

When the rupture and the disharmony are overcome, and one of the presumptive possibilities is confirmed and the other rejected, there is the confirmatory "yes" (or, "no"), which is again a modalisation of the original certainty: "It is indeed, in fact, really so." The original unmodified "being" acquires the character of "really, so" (Hua XI, 230).[15]

(c) Within pre-predicative experience there is room for the origin of yet another concept of possibility. Every perception takes place within a horizon of familiarity. This horizon, both internal and external, gives rise to expectations, which constitute possibilities that are not mere phantasy-possibilities, but are rather supported by the original positional consciousness of validity (Hua VIII, 148). What is intentionally pre-delineated in the horizon of a perception is not possible but certain. This pre-delineation itself -- that, e.g., the other, unseen side of the perceived physical object has some colour, some tactile quality -- is, as Husserl characterises it at one place, an "unbestimmt allgemeine" (Hua XI, 40). But this generality has an extension of free variability. What falls within it has an "open possibility". On the basis of the present perception, nothing counts in favour or against. All open possibilities are equally possible. It is obviously otherwise with the presumptive possibilities, where not only each alternative is specifically intended, but something counts in favour of one as against another.

(d) Or, consider a thing that is there before me in perception. Just as this *fact* of perception may be "transformed", in phantasy, into a *possible* perception, so also may its object. In fact, just as every intentional experience is permeated with potentialities which give rise to open possibilities, so also the entire sphere of consciousness is characterised by the difference between actuality and phantasy. Every actual consciousness can be "transformed" into an as-if consciousness and every actual object into an as-if object. These possible perceptions, possible imaginings -- as well as possible objects -- may provide the basis for bringing to intuitive givenness the pure eidos 'perception', the pure eidos 'imagination', as well as the pure eidos 'object', whose ideal extensions are made up of all appropriate possible perceptions, imaginings and objects.[16] It should be noted that this as-if transformation can only be purely *theoretically* motivated. These as-if possibilities as well as the *eide* whose ideal extensions consist in such possibilities are often called by Husserl, certainly in the later writings, 'pure possibilities' understood as pure imaginables. Phantasy, as an imaginative act, is constitutive of such possibilities.

(e) We have now followed a whole series of forms of consciousness in which different forms of possibilities are given. By virtue of the fact that consciousness can always objectify any of its achievements, "It is possible that ... " may be objectified as "the possibility of ... ", whereby an objectivity of a higher order is constituted out of the appropriate modalised forms of consciousness and their objects.

Earlier I had remarked that all three major concepts of possibility -- the Leibnizian, the Aristotelian and the Kantian -- are to be found in Husserl. It is time now briefly to indicate in what precise form each of these is to be found.

The Leibnizian concept is on the surface of a large part of the exposition in the first part of this paper. The concept of pure possibilities, and of many compossible worlds of which ours is one -- all these are admittedly appropriated by Husserl, without no doubt the Leibnizian theodicy and the concept of God's intellect as the original home of those possibilities. Only one compossible world, Husserl agrees, is actual. In §60 of the *Cartesian Meditations*, there is an attempt to interpret this thesis. There cannot be two communities of egos, there can exist only one such community. Any two such communities cannot be absolutely isolated. If two such are imagined by me, then, as the "constitutive primal monad relative to them", they are in communion through me. Like two possible variants of my ego, two closed worlds are incompossible. The major departure from Leibniz lies in this that pure possibilities, instead of belonging to God's intellect, are constituted in the life of transcendental subjectivity. The idea of 'compossibility' receives an appropriate interpretation within transcendental phenomenology: a world relates to a monadological intersubjectivity, and every monadological intersubjectivity is oriented around one primal monad, my ego. It can also be yours.

The Kantian thesis is partly retained, but also, in part, abandoned. Of all the Kantian categories, it is the categories of modality which do not constitute the object in the strict sense. Once the object is there, the modalities are concerned with "how the object, together with all its determinations, is related to understanding and its empirical employment, to empirical judgment, and to reason in its application to experience."[17] For Husserl also, the predicates "actual" and "possible" do not determine the object. "A is actual" and "A is possible" are not determinative propositions.[18] The former means, for example, "I do not imagine A", the latter "A is imaginable". The "modes of being" are correlates of thetic act qualities. They must be distinguished, even if not separable, from the objective *Sinn*. The Kantian position is modified, in so far as Husserl separates 'actuality' from 'existence', and further in so far as the peculiar Kantian concept of forms of understanding is not there in Husserl. Furthermore, Husserl continues to hold the Leibnizian thesis that more is possible than actual, a thesis Kant explicitly rejects.[19]

The Aristotelian concept of potency is appropriated into the thesis that the actual is constituted by pre-delineated potentialities and that full actuality as the correlate of perfect evidence is *progressively* achieved by gradual fulfillment of these potentialities. And yet this has nothing of the correlative Aristotelian concepts of matter and form. Furthermore, actuality is not the mere end state of the process, but at every stage actuality and potentiality seem to interpenetrate.

5. "PRACTICAL POSSIBILITY"

Underlying this widely ramified theory of possibility, there is, for Husserl, one most fundamental, and from the point of view of constitutive phenomenology, primitive possibility-consciousness -- the practical possibility symbolized by the expression "I can". The distinction between logical possibility and practical possibility is formulated in *Ideen* II thus: Both are derived from neutrality modification, the former from neutrality modification of intuitive representation (into the mode of as-if non-actuality), the latter from neutrality modification of acting into a quasi-acting. "What I can, am able to, that of which I know myself as capable, ... that is a practical possibility. I can 'decide' only from amongst practical possibilities; only a practical possibility can be the theme of my will" (Hua IV, 258ff). Note that the last 'can' is a theoretical 'can'.

It seems to me that in this account of the distinction the symmetry is misleading. If "transforming" an actuality into an as-if object generates (one sort of) logical possibility, there is an "I can"-consciousness which is not yet a neutralised consciousness of "I can", which is rather involved in, and an implicate of, some modes of acting. It is this original "I can"-consciousness, and not the one "represented" in a phantasy-modification of acting (as in "I can move my hand" said on the basis of mere representation *of* an action), which is in my view the primal source of practical possibility. Of course, like all modalities, this also may be reflectively thematised and objectified.

Once this point is grasped, we can understand the empty horizons belonging to any experience as "practical horizon" including a "system of possibilities for practical intervention". One may even understand the pre-delineated possibilities of fulfillment as practical possibilities (Hua VII, 275). As a matter of fact, one may want to assimilate all doxic-logical possibilities to the practical "I can phantasise" (Hua IX, 205). *The primacy of the practical* receives a still stronger support from the following consideration: since positing any object, real or ideal, implies the possibility -- egological as well as inter-subjective -- of reiterating certain confirmatory and evidence-producing processes (such as possible perceptions in the sense of perceptions I *would have* were I to walk around the object), "without such 'possibilities' there would be for us no *fixed and abiding* being, no real and no ideal world".[20] This especially holds good of the pure possibilities. They are ideal objects, their being beyond time consists in being correlates of "free produceability and reproduceability at all times".[21]

However, there is something dissatisfying in subsuming all acts that generate possibilities under the concept of practical possibility. It is true that whether it is pure possibility in the strict sense, or pure possibility in the sense of an imagined variation of an actuality, or as-if transformation of an actuality, or open possibilities pre-delineated in the horizon of a perceptual experience, or motivated possibilities arising out of obstruction of perceptual intentions -- in every case, we have an "I can"-form of consciousness at the source of the constitution of the modality. But it is also arguable that not all "I can" is "practical" in the strict sense. Practical in the paradigmatic sense for Husserl is, of course, the kinaesthetic consciousness "I can move myself". Only in a much less appropriate sense, one can characterise "I can imagine" as practical. This is not to dispute the contention, well formulated by Landgrebe, that possibilities, in their original manner, are what can take place precisely because I bring them about, that the root of our consciousness of possibility is in the "*Vermöglichkeiten*" of our dispositional capacities.[22] What I am disputing is that all these *Vermöglichkeiten* are bound up with the awareness "I can move myself" or its like. The latter may be the most primitive awareness of possibility on which the distinction between the actual and the possible is founded, but there would appear to be higher forms of "I can"-consciousness which are not tied to corporeality in the manner kinaesthesis is. We are still unclear, for example, as to how "I can imagine" is related to "I can move". It is possible to subsume the theoretical "I can" under the practical by speaking of "theoretical practice", and Husserl sometimes does this. But in doing so we end up by extending the meaning of "practical" in a manner that is surely not illegitimate, but possibly not enlightening. A paradigmatic example of the theoretical "I can" is the constitution of the possibility of the world's dissolving into nothingness. This is not a motivated possibility for which anything in experience speaks. There is no real possibility of its non-being. I cannot imagine it. Yet there is an open possibility that this harmonious unitary structure of world-experience dissolves itself (Hua VIII, 45-55). *This* open possibility -- note that this is a different sense of "openness" than that in which the potentialities pre-delineated in the horizon of a perceptual experience are open -- is not constituted by the practical "I can", but entirely by a theoretical "I can conceive of it".

To sum up: we have followed two movements, in opposed directions, one from above, i.e. from abstract pure possibilities to the actual, and the other from below, i.e. from the actual to the possibilities of different degrees of remoteness. The former is the path of ontology, the latter the path of phenomenology. The latter movement requires the appropriate constitutive acts of consciousness. But underlying the actual performance of those acts is the dispositional "I can" consciousness which, I have insisted, is of two radically different sorts: practical and theoretical. Each may subsequently be objectified, yielding possibilities as entities.

NOTES

[1] Compare my "Intentionality and Possbile World: Husserl and Hintikka," in H.L. Dreyfus, ed., *Husserl, Intentionality and Cognitive Science* (Cambridge: M.I.T. Press, 1982), pp. 233-251.

[2] E. Husserl, *Logical Investigations*, trans. J.N. Finlay (New York: Humanities Press, 1970), p. 345. Hereafter cited as *LI.*

[3] E. Husserl, *Cartesian Meditations*, trans. D. Cairns (The Hague: Martinus Nijhoff, 1960), § 19. Hereafter cited as *CM.*

[4] *CM*, §27.

[5] E. Husserl, *Paris Lectures*, trans. P. Koestenbaum (The Hague: Martinus Nijhoff, 1964), p. 33.

[6] A. Gurwitsch, *Leibniz: Philosophie des Panlogismus* (Berlin: Walter de Gruyter, 1974). The English translation is mine.

[7] *LI* II, pp. 751-753.

[8] This coincidence of modes has been emphasised by Nicolai Hartmann in his *Möglichkeit und Wirklichkeit* (Berlin: Walter de Gruyter, 1949).

[9] *CM*, § 60.

[10] Nicolai Hartmann, op.cit.

[11] Compare my "Remarks on Nicolai Hartmann's Modal Doctrine," *Kant Studien* 54 (1963), pp. 181-187. Reprinted in *Phenomenology and Ontology* (The Hague: Martinus Nijhoff, 1970).

[12] I. Kant, *Critique of Pure Reason*, A 219 = B 266.

[13] E. Husserl, *Experience and Judgment*, trans J.S. Churchill and K. Ameriks (Evanston: Northwestern Press, 1973), p. 100. Hereafter cited as *EJ*.

[14] *EJ*, p. 95.

[15] Cf. *EJ*, p. 100.

[16] *CM*, §§ 25 and 34.

[17] *Critique of Pure Reason*, A 219 = B 266.

[18] *EJ*, §§ 75-76.

[19] *Critique of Pure Reason*, A 230-1 = B 282-4.

[20] *CM*, § 27.

[21] *CM*, § 55.

[22] L. Landgrebe, "The Phenomenological Concept of Experience," *Philosophy and Phenomenological Research* 34 (1973), esp. p. 10.

10. PHENOMENOLOGY AND THE MODALITIES *

My investigations into the nature of the modalities in Husserl's philosophy were first undertaken under the influence of the pioneering work of Jaakko Hintikka. I looked into the nature of Husserl's various concepts of possibility[1], and subjected the possible-worlds interpretation of Husserl to the criticism that it involved a naive ontological reification.[2] Hintikka and Harvey[3], in response to my criticisms, and Seebohm on his own, have supplemented the possible world theory with an account — culled from Husserl's writings — of the *constitutive processes* of modalization, thereby introducing the needed genetic underpinning, unavoidable from the phenomenological point of view, for the static-descriptive account of the earlier work (of Hintikka). All these have set the framework for study of the other modalities in phenomenology. In this paper, I will retrace some of the steps taken earlier, with regard to 'possibility'.

I

The concept of possibility plays a central role in the phenomenological philosophies. This is true not only of Husserl, but also of Heidegger and Sartre. Since Husserl's views will be discussed in the larger part of this paper, let me begin by referring to Heidegger and Sartre. First, Heidegger:

> "Dasein ist nicht ein Vorhandenes, das als Zugabe noch besitzt, etwas zu können, sondern es ist primär Möglichsein".[4]

This possibility, we are then told, is not "the empty logical possibility". The distinction between this existential possibility and mere logical possibility is thus drawn by Heidegger:

> "Als modale Kategorie der Vorhandenheit bedeutet Möglichkeit das *noch* nicht Wirkliche und das *nicht jemals* Notwendige. Sie charakterisiert das *nur* mögliche. Sie ist ontologisch niedriger als Wirklichkeit und Notwendigkeit. Die Möglichkeit als Existenzial dagegen ist die ursprünglichste und letzte positive ontologische Bestimmtheit des Daseins."[5]

---

* Originally appeared in *Acta Philosophica Fennica*, 49, 1990, 110-122.

'Comprehension' (*Verstehen*), as an existential moment of *Dasein*, is entirely made up of possibilities. The possibilities that are projected by *Dasein*, are not thereby thematically apprehended, for such thematization would, Heidegger writes, deprive the projected of its possibility-character. *Dasein*, as such projection of possibilities of being, is always more than what it in fact (*tatsächlich*) is, but it is not more than its own facticity (*Faktizität*), for its facticity precisely consists in its possibilities of being (*Seinkönnen*)[6].

Later in *Vom Wesen des Grundes*, Heidegger writes:

> "Im stiftenden Gründen als dem Entwurf von *Möglichkeiten seiner selbst* liegt nun aber, dass sich das Dasein darin jeweils *überschwingt*. Der Entwurf von Möglichkeiten ist seinem Wesen nach jeweils reicher als der im Entwerfenden schon ruhende Besitz. Ein solcher eignet aber dem Dasein...damit sind dem Dasein bereits bestimmte andere Möglichkeiten — und zwar lediglich durch seine eigene andere Faktizität — *entzogen*. Aber gerade dieser in der Eingenommenheit vom Seienden beschlossene *Entzug* gewisser Möglichkeiten seines In-der-Welt-sein-könnens bringt erst die wirklich ergreifbaren Möglichkeiten des Weltentwurfs dem Dasein als seine Welt *entgegen*."[7]

In this account, there are two important theses that we need to bear in mind. First, human existence is not like that of a stone or a plant, it consists in projecting possibilities on the basis of its facticity, not empty logical possibilities, but possibilities of its own being. Secondly, one must distinguish between the original and authentic projection of possibilities, and a derived, subsequent thematization of those possibilities; the former alone faces the possibilities as possibilities, the latter transforms them into given, presented, objects.

Let me now briefly refer to Sartre. Sartre recognizes that possibility, in *the subjective sense* in which it does not belong to being, but is conferred by a subject upon it, and exists only in relation to our ignorance, must be distinguished from *ontological* possibility which belongs to man, to a for-itself, as *its* possibility: it is the possibility which it is, which it has to be. In the latter case, Sartre says, a being is the ground of its own possibilities, it is their foundation.[8] What about the Leibnizian possibilities as existing in God's understanding?

According to Sartre, as for Heidegger — these possibilities, as thoughts, *as represented*, cease to be possibilities, they loose their nature as possibilities and are reduced to the subjective being of representation[9]. Thus, he writes:

> "If we consider possibles in the heart of the divine understanding as the content of the divine thought, they become purely and simply concrete *representations.*"[10]

Genuine possibility, therefore, cannot be subjective. It must be ontological, and can belong not to being-in-itself (which is what it is), but only to being-for-itself (which is what it is not). "Possibilities" come into the world through human reality[11]. The in-itself cannot "have" possibilities. It is easy to recognize that this view of Sartre is very close to Heidegger's.

The views of Heidegger and Sartre clearly suggest that a genuinely phenomenological concern with possibility must be fundamentally *opposed to* a theory of possible worlds as commonly construed. There is also no doubt that the Heideggerean position derives from a certain construal of the Husserlian understanding of intentionality. And, yet, as has been shown by Hintikka, and following him by David Smith and Ron McIntyre, Husserlian theory of intentionality does have elements which permit to be developed into a possible-worlds theory. Let me then, with this complex problematic in mind, turn to Husserl.

II

There are two places in Husserl's thought, which permit a natural connection to a possible worlds theory. First, there is the essentialistic ontology, laid down in broad outlines, in chapter 1 of the *Ideen,* Bd I. The second place is the idea of 'horizon', which has been well exploited in recent work on Husserl. The former has more natural ties to Leibnizian ontology. Are the two connected in any interesting manner?

The essentialism may be formulated in the following four theses:

1."To each essence there corresponds a series of possible individuals as its factual instancings."[12]

2."Individual Being of every kind is, to speak generally, 'accidental'. It is so-and-so, but essentially it could be other than what it is."[13]

3."... possibilities themselves are ideal objects. Possibilities can as little be found in the real world, as can numbers in general, or triangles in general."[14]

4."What is here called possibility (eidetic existence) is ... absolutely necessary possibility, an absolutely firm joint in the absolutely firm structure of an eidetic system. The goal of enquiry is to know this system scientifically..."[15]

Husserl speaks of 'pure possibility' in two different contexts: first, pure possibilities are often said to be (merely) logical possibilities, and something is logically possible when it is consistent, does not involve self-contradiction, etc. etc. Thus, a 'golden mountain' is logically possible, a 'round-square' or a married bachelor is not.

What about 'something that is colored but not extended' or 'a perception that is not perception *of* an object' or 'a hope that is not a hope for something'? These are not possibilities, not logical impossibilities, but what Husserl calls *eidetic* impossibilities.

I think, Husserl held that not only are essential, eidetic modalities (possibility-impossibility, necessity-contingency) irreducible to the logical modalities, but, on the contrary the logical modalities are cases of essential modalities. What he meant is perhaps something like this: if a compound concept does not involve material or contentual concepts, but only formal-logical concepts and variables (such as '*A* which is *A*', '*A* which is non-A' or '*A* which is *B*'), then its consistency or lack of it, follows from the *essence* of the logical as such. That is why, even if these are either merely logical necessities, contradictions or possibilities, they are *also* essential, eidetic necessities, contradictions or possibilities inasmuch as they follow from the essence ('formal essence', that is) of the logical as such. However, ordinarily when one speaks of essential modalities, one has in mind not these merely logical modalities, but modalities pertaining to material (as opposed to formal) essences. Now one also says of a compound *concept* that contains non-logical, material concepts that it is consistent or self-contradictory or necessary (and so, the correlative *object* possible, impossible or necessary), one may think what one has on hand are logical modalities inasmuch as one is able to demonstrate them logically, but one can do so only if one already has at one's disposal definitions of the concepts involved, and such definitions, if they are not sheer conventions, have to be grounded in the essences of the matter at hand. Thus, for Husserl, logical modalities are subsets of eidetic modalities.

If *P* is a non-compound essence, then *P* is a pure possibility - by which is also meant, besides saying that *P* is an essence, that there are possible individual *P*'s as its instances. So one says, it is a pure possibility that *P*'s obtain, a possibility that follows from *P* being an essence.

If *P* is, as is more often the case than not, a compound essence, containing two component essences *A* and *B*, then *P* is a pure possibility only if *A* and *B* are consistent. (*If A* is a simple essence, it is consistent with itself and so a pure possibility.) Whether *A* and *B* are compatible or not, depends upon what sort of unification is under question. They may be compatible in relation to a whole of the type *W*, and not compatible in relation to a whole of the type *W'*. What, then, is a pure possibility is the compound essence *W*(*A B*). This is the same as saying *W*(*A B*) is a compound essence, *W'*(*A B*) is not.

We need to know what is the sense of 'consistency' or 'compatibility' (or, their negations) that is appropriate here. The formal-logical idea of 'freedom from logical contradiction' holds of logical forms; compound logical forms, if consistent, are pure possibilities. We are indeed seeking for a notion of pure possibility of which logical possibility — as said before — is a species. When thus pressed for a non-logical sense of 'compatibility' Husserl does indeed come up with one: A compound concept is consistent if the meaning intention sustaining it has a 'fulfilling sense'; in other words, if the partial intuitions of the component meanings can be combined into a total intuition. In that case, the compound objectivity is *conceivable,* the essence has an ideal *existence,* it is a pure possibility. These *three* locutions amount to the same.

It has been noted that if *P* is an essence, i.e. is a pure possibility, it is also a pure possibility that *P*'s obtain. Now this equivocation in the sense of 'pure possibility' needs to be cleared up.

An essence is either dependent or independent. A dependent essence — which Husserl calls *'abstractum'* - such as *"red circle"* cannot exist by itself. An individual red circle — which instantiates the essence *red circle* — has necessarily more determinations than merely being red and being a circle. Such an abstract, i.e. dependent essence is not capable of *real* existence (although, it may be said to have its own, i.e. ideal actuality). Pure possibilities in this sense have ideal actuality, they are 'discovered', and truths grounded in them (i.e. concerning relations of dependence of, or founding-founded between, two such dependent abstract essences) are strictly universal and necessary.

When however such an abstract essence *P* is instantiated in reality, what we have is a real individual *P* whose essence consists of the essence *P* supplemented by a host of other determinations. This complex nexus of essences [*P*, *Q*, *R*...] — consisting of essences which need each other for supplementation in order to be *really* instantiated, and each essence specified down to the lowest *infima species* (e.g. '*red*' to a *specific shade of red*, *circle* to circle with a given radius and so on and so forth) — is capable of having real existence, without any further supplementation, it is an independent essence, a *concretum.* It is also a pure possibility, an ideal actuality — although it is the complete essence of a possible reality, the individual it determines is a possible, even if not actual, reality.

Contrast *vague Adam* with a *possible variant of Adam* or *triangle* with *the triangle drawn on the blackboard before me. The triangle...* is a concrete essence which is the essence of an actual real triangle. Both *vague Adam* and *triangle* are abstract essences incapable of being realized, without further supplementation. *A possible Adam,* i.e. actual Adam but with some variations, is a concrete essence which can be realized in a possible individual.

If we assume that a world can consist only of real, determinate, individuals, and since not all concrete essences though capable of real instantiation are not in fact so instantiated (as, e.g., not all variations of Adam are *realized),* there are more possible individuals than are real (and so more possible worlds than the actual one). But these possible individual worlds are not a chaotic heap, their complex essences must be subject to strictly universal and necessary laws that determine the relations of mutual dependence, compatibility/incompatibility, founded/founding amongst essences, laws that constitute the grid of what Husserl calls in the text quoted above "the absolutely firm structure of an eidetic system" which to know scientifically is the goal of all knowledge.

There are several very difficult questions with regard to this essentialism which I will presently skip over, and ask only *one* of them just because it shall provide a nice transition to the second aspect of Husserl's thought mentioned earlier. The question is, even if there are essences, how are they available to us? How do we ever get to discover them? Here traditional phenomenology has taken recourse to the method of eidetic or imaginative variation. If one is to determine the essence of art works, more specifically of literary works of art, one needs to start with some exemplar (which need not be an actual case like a play of Shakespeare, but may be an imagined one) and then continue imagining possible variations upon it until such time as one would *not* want to apply that concept. With this, one, at one stroke, reaches the limits of its application as also the essence sought for. Again, there are various aspects of this method which call for discussion, but there is only one that I intend to draw your attention to now: this is the use made of the concept of possible variations upon the exemplar chosen. The method is no different from when, looking for the essence of Socrates, one imagines possible variations of the actual Socrates, and then continues to ask, shall we call this variant Socrates? (For example, if Socrates were not snub-nosed but had all the other features that in fact he has, we would not hesitate to call him Socrates, but what would we want to do if he had all the other properties save his wisdom?) In effect, we want 'Socrates' to refer to the same individual in many possible worlds — the *same in essence,* though not in all respects. (This much, I wish to add, is compatible with the choice of any of the competing theories available on this matter.) What could motivate such imagining, phantasizing, possible variations? Husserl writes:

> "It is an empty possibility that this writing-desk here on its underside, which is at present invisible to me, rests on ten legs instead of on four legs, as is really the case. This four-ness, on the contrary, is a motivated possibility in respect of the definite perception which I am just enjoying."[16]

The precise meaning of 'motivation', and so of 'motivated possibility' needs to be determined, but I will not attempt it here. Suffice it to note that nothing in my perception of tables speaks in favor of the imaginative variation just spoken of.

Essences are thus given in, and, in one sense, constituted by such acts of unmotivated *imaginative* variation.

On the other hand, every perception by virtue of its implicit horizonal character points to, i.e. *motivates,* other possibilities of further unfoldment of the perceptual process. This thesis about 'perception' is but a special case of a general thesis: "All actual experience refers beyond itself to possible experiences, which themselves again point to new possible experiences, and so on *ad infinitum*".[17] We are also told that the horizon of an act is "predelineated" in the act, but it is also predelineated in its "determinate indeterminateness" (e.g. the backside of that wall must be colored, though no particular color is determined). This openness is an element included in the consciousness itself as its horizon. These predelineated "potentialities" are not mere empty logical possibilities, but are motivated by the nexus of experience, by *Erfahrungszusammenhänge.* A possible world interpretation of Husserlian intentionality is not very far from this thesis: such an interpretation, to quote from Smith and McIntyre,

> "associates with each act of consciousness both a noema and a "horizon" of possible worlds compatible with what the *Sinn* of this noema (together with the act's background *Sinne)* prescribes."[18]

A further extension would yield the thesis that "an act is not directed towards an object, or about an individual, *simpliciter* but only in, or relative to, a possible world".

How then do that essentialism and this horizon theory hang together? To determine the essence of a material thing is to carve out a *necessary* structure *within which* there are all the possibilities of variations of any material thing. If I perceive something as a material thing, I cannot be perceiving it as possessing features that are ruled out by virtue of its essence. The essence thus lays down *the outermost horizon* of possibilities for an object that I may be experiencing. Within this outermost horizon, there is, for any given experience, an *inner horizon* of possibilities, that are "predelineated" in the course of experience, i.e. which are motivated by the course experience has run: this is a subset of all the possibilities that the essence allows for. The two theories — that essentialism and this horizon theory, the two "places" where the concept of possibility intrudes into phenomenology — are thus of a piece: they both seek to explicate the restrictions on the *available* range of possibilities.

III

How does the possible worlds theory fare from a phenomenological perspective? On this large and intricate theme, I will here offer only a few general remarks.

In an earlier paper devoted to this topic, I had remarked that the possible worlds theory is doomed to "the naïveté that does not look into their constitutive source by tracing back the process of 'modalization' as it operates both in prepredicative and predicative experience, and the activity of thematisation which *objectifies* possibilities into a new class of entities." This is precisely what Harvey and Hintikka now do.

Another anxiety which phenomenologists are apt to express, when faced with a possible worlds theory, is that an infinite set of possible worlds could not possibly form the content of one's consciousness — whether it be in understanding utterances or in perceiving a material object. Consequently, postulating such a set as "implicitly" present in an experience, is tantamount to overstepping the limits of phenomenology. What is needed, to assuage this anxiety, is some principle of limitation upon this infinite set, so that one is fair to the limitations of the explicit and implicit contents of one's consciousness. Hintikka's earlier limitation that we admit only those possible worlds which are *compatible* with one's perceptual content is too restrictive, for this entails that in the unfolding of the perceptual process the content at any given stage will never come to be contradicted.

Hintikka, in some of his later work, recognizes this problem, and concedes that an unlimited possible world analysis of meanings is an illusion. Accepting my earlier criticism as legitimate, Harvey and Hintikka suggest that the idea of possible worlds must be conceptually tied to a "home world" along with a "principle of production" located within that world. This principle of production of possibilities is the knowing, thinking subject. Harvey and Hintikka then go on to show, closely following Husserl's genealogy of logic, how various sorts of possibilities are "spawned, from the subject's encounter with the actual world" — into the details of which we need not enter at present.

A third worry concerns the locution *"all* possible worlds". Here essentialism and the theory of horizon seem to have different stories to tell. Essentialism appears to be able to close off the domain of possibility, so that one who knows the essence knows *all* that is possible. As a matter of fact, he knows what is ruled out of possibility, he does not know *each* possibility. Give him one and he will recognize it. The theory of horizon lets the possibilities unfold in temporal succession, all these possibilities never are together so as to form a totality. They cannot be given, even to an omniscient being, as a totality. Thus, the thesis of *all* possible worlds of which only one is actual, or actual for us, is a *thesis from God's point of view.* Nothing in experience warrants this reification and *totalization.* The thought of it would remain empty — merely symbolic.

Finally, it must be kept in view that any possible world derives its sense from its being conceivable by me and the intersubjective community to which I belong. If the actual world is the correlate of the actual course of my experiencings, any possible world would be the correlate of some variations of my or our empirical consciousness. Now although the course of my experience may run differently than it does, or might have been different from what it is, it is, in any case, conceivable only as *my* experience. I can imagine only as much variation of my experience as is compatible with its being mine. This 'I', the ultimate point of reference of all conceiving of possibilities, is necessarily the actual I. There is no *possible* I. What I am suggesting is that all possibilities refer back to the *actual* thinking ego. This is the meaning of the following text from Husserl's *Cartesian Meditations:*

> "As imagined by me, each of them [of *two* intersubjectivities] is in necessary communion with me ... as the constitutive primal monad relative to them." (60)

IV

There is still another aspect of Husserl's theory of possibility to which I have to refer. This may be brought out by asking, whether in the two terms, 'a possible guest' and 'the guest first to arrive', the word 'guest' is being determined by the two qualifiers in the same manner. It is easy to see that a possible guest is not a guest, nor is he/she a kind of guest. A possible philosopher is not a philosopher — nor a kind of philosopher, in the same sense in which a white horse is a horse and a kind of horse. Using Brentano's distinction between determining and modifying predicates, one can say that 'possible' is not a determining predicate (as 'red' in 'red flower' is), but a modifying one (as 'dead' in 'dead man' is, to take one of Brentano's examples). I would even go further, and distinguish between the sense in which 'dead'/'living' are 'modifying' predicates, and the sense in which 'possible', 'actual' and 'necessary' are. This distinction can be brought out by asking, of what are these latter predicated? In the tradition of thinking about modalities, there are two standard answers to this question. According to one, they are predicated of things, according to the other, they are predicated of thoughts (concepts, propositions, sentences). They are either de re or de dicto predicates. Husserl's view seems to have been that they are, originally, predicates of senses or meanings or noemata.

Let me quote some sentences from *Experience and Judgment*:

"In the natural attitude, there is at first (prior to reflection) no predicate 'actual,' no genus 'actuality.'"[19] 'A is actual' contrasts with 'A is a fiction/non-actual'. "The object, e.g., the house, acquires no 'determination' in the proper sense, no further determination in the way of an explication ('The house is red'), if the object is designated as an actuality."[20] The house does not belong to the class of objects which are actualities, as if there were a class of objects which are not actualities — every class of objects is a class of actualities.

As an object of experience it is actual; as an object of imagination, it is not. "The same" object in one case is actual, in another imaginary. What is this "the same"? There is a common core of sense, but there are also noematic predicates in the one case, which are not present in the other: these are what Husserl elsewhere calls "thetic" predicates. One can also call this "noematic core" (not taking into account the thetic predicates) "the individual essense" of the same object that is actual but could have been non-actual, or is merely imagined but could have been actual.

The matter becomes more muddled when we ask of two imaginary objects, if they are "the same" — whether, the Gretel of one fairy tale and the Gretel of another are the same Gretel. Husserl, in one text, concedes that we can only stipulate if they are or not. For, in the strict sense, individuation and identity of the individual are possible only within the world of experience on the basis of absolute temporal position.

To sum up: the complexity of the account I have here presented is derived from looking at the modalities from three different perspectives: from the perspective of an ontological essentialism, from the perspective of a temporally unfolding horizonal character of experiencings, and from the perspective of the noematic meaning-contents of intentional acts. To see the inner connection between these perspectives, we must be able to see the connection between these three: essence, experience and meaning/noema.

I said earlier that for Husserl, essential possibility is the key concept, 'logical possibility' is subordinated to it. A notion of 'essential possibility' that is irreducible to the merely logical, is his original contribution. One can say the same of 'necessity'. It would be possible to show that the two concepts — 'eidetic possibility' and 'eidetic necessity' — belong together, are two sides of the same and stand or fall together. It is one of the deficiencies of the existential phenomenologists that they took over — as we saw at the beginning — the idea of an essential possibility, but did not want to have anything to do with eidetic necessity.

Now, essential necessity is neither empirical-causal necessity nor logical necessity. Examples of essential necessity that are often adduced in phenomenological literature are:

(1) All that is colored is extended;

(2) Consciousness is (always) consciousness-of;

(3) Outer perception is perspectival;

or, to take one from Heidegger:

(4) Dasein is being towards death.

I will not here go into the various attempts made in the literature to show that (1)–(3) are really logical truths. But all these attempts presuppose appropriate definitions of the non-logical terms involved. Such definitions if they are to be non-arbitrary, must state the corresponding essences.

NOTES

[1] Mohanty, J.N., "Husserl on 'Possibility'", *Husserl Studies*, vol. 1, No. 1, 1984.

[2] Mohanty, J.N., "Intentionality and Possible Worlds: Husserl and Hintikka", in Hubert L. Dreyfus ed., *Husserl, Intentionality and Cognitive Science*, M.I.T. Press, Cambridge, 1982, pp. 233-255.

[3] Harvey, C. and Hintikka, J., *Modalization and Modalities,* forthcoming.

[4] Heidegger, M., *Sein und Zeit*, 7.Aufl., Max Niemeyer, Tübingen, 1953, p.143.

[5] Ibid, pp.143-144.

[6] Ibid, p.145, also see p.262.

[7] Heidegger, M., *Vom Wesen des Grundes*, Vittorio Klostermann, Frankfurt, 1949, p.43.

[8] Sartre, J-P., *Being and Nothingness*, tr. Hazel E. Barnes, Methuen & Co., London, 1957, p.81.

[9] Ibid, p.97.

[10] Ibid, p.98.

[11] Ibid, p.97.

[12] Husserl, E., *Ideas* I, tr. Boyce Gibson, Macmillan, New York, 1931, p.55.

[13] Ibid, p.47.

[14] Husserl, E., *Logical Investigations*, tr., J.N. Findley, Humanities Press, New York, 1970, vol. I., p.345.

[15] *Ideas* I, p.347.

[16] Ibid, p.361.

[17] Ibid, p.47.

[18] Smith, D. and McIntyre, R., *Husserl and Intentionality. A Study of Mind, Meaning and Language*, Reidel, Dordrecht, 1982, p.311.

[19] Husserl, E., *Experience and Judgment*, tr. James S. Churchill, Northwestern University Press, Evanston, 1973, p.298.

[20] Ibid, pp.301-302.

11. HUSSERL'S 'LOGIC OF TRUTH'*

I. PRELIMINARY CONSIDERATIONS

Husserl's distinction between three strata of formal logic — pure logical grammar, logic of consequence and logic of truth — has been well recognized and much discussed in the secondary literature,[1] and may well have exercised some influence on the actual development of modern logic. But very little has been done to explain what he could have meant by 'logic of truth' — much less to further develop that idea. I want this paper to be a contribution to that theme.

It would be best to begin with some well-known features of the distinction between 'logic of consequence' and 'logic of truth'. The first of these is that logic of consequence (or what may also be called, logic of non-contradiction) investigates into the essential laws of determining analytic inclusion and exclusion of propositions, the analytic consistency of propositions, i.e., with their mutual compatibility or lack of it. The basic concepts of this part of formal logic are 'analytic consequence' and 'contradiction'. 'Truth' and 'falsity', along with their modalities, do not belong to its fundamental concepts. The logic of truth, as contrasted with the logic of consequence, precisely deals with the concepts of truth and falsity (and their modal derivatives) and the general laws pertaining to them. With regard to the fundamental logical principles, Husserl shows how their formulations change from one stratum of logic to another. The principle of non-contradiction and the principle of excluded middle, for example, have to be formulated in the logic of consequence without making use of the concepts of truth and falsity, and in logic of truth by making use of them.[2] I will return to this last claim at a later point in this essay.

Secondly, the very concept of 'judgment' is different in one case and in the other. In logic of consequence, a judgment belongs to its domain in case it satisfies the laws of pure logical grammar, i.e., is syntactically well-formed. In

---

* Originally in *Mind, Meaning and Mathematics*, L. Haaparanta, ed. Kluwer Academic Publishers, 1994, pp. 141-160

logic of truth, a judgment must also satisfy the requirements of semantic appropriateness and semantic compatibility. Husserl assigns to each a different concept of evidence. To the judgment as it is taken up in logic of consequence, he assigns the evidence of distinctness (*Deutlichkeit*), to the judgment as it is taken up in logic of truth, he assigns the evidence of clarity.[3]

Thirdly, logic of truth takes into account, while logic of consequence does not, the natural striving of one's intentional, cognitive life to reach truth. Thus the logic of consequence regards judgments as ideal propositions and enquires into the relations of compatibility, incompatibility and implication amongst them, without taking into account their actual truth or falsity in any given cognitive situation. Traditional formal logic, Husserl insists, has not been a pure logic of non-contradiction, it did not clearly recognize the distinction of logic of non-contradiction from the logic of truth[4] — a distinction which Husserl claims to be his original discovery.[5]

Of these three features of Husserl's distinction, the second, i.e., the idea of different sorts of evidence has been discussed and elaborated at length by G. Heffernan.[6] The third, i.e., locating a stratum of formal logic within the overall context of epistemic striving after truth, has been the focus of Sokolowski's work.[7] In spite of these and other researchers, the idea of logic of truth remains in the dark and open to various interpretations.

II. AN EXCURSUS INTO KANT

Kant's distinction between formal logic and transcendental logic appears to anticipate Husserl's distinction between consequence logic and truth logic, especially since Kant also calls the Analytic part of his transcendental logic 'logic of truth'.[8] Husserl's characterization of his distinction is also not entirely dissimilar to Kant's. Formal logic — or general logic as Kant calls it — abstracts from all contents of thought and deals with the mere form apart from "all relation of knowledge to the object." It considers "only the logical form in the relation of any knowledge to other knowledge."[9] In this last phrase, replace 'knowledge' by 'judgment,' and we get part of Husserl's characterization. Transcendental logic, however, on Kant's account, abstracts only from all empirical contents, but "would treat of the origin of the modes in which we know objects, in so far as that origin cannot be attributed to objects" (i.e., is *a priori*).[10] Transcendental logic, then, would be a theory of our *a priori* knowledge of objects. It is epistemic logic. Thus the table of the categories belongs to transcendental logic, as also the question of their *a priori* objective reality. This is not quite the same as Husserl's logic of truth. Nevertheless, Husserl's logic of truth is also, in some sense, epistemic logic, although Husserl still wants it to be a formal logic of truth. In other words, whereas Kant contrasts transcendental logic with formal logic, for Husserl the logic of truth is a stratum within formal logic. We still need to look into the question, how Husserl's transcendental logic relates to his logic of truth on the one hand and to Kant's transcendental logic on the other.

Kant, in the very same introductory paragraphs of the chapter on "Transcendental Logic", goes on to raise a fundamental objection against the possibility of a logic of truth *understood in a certain manner*. He says that a general criterion of truth, if any such is advanced, cannot, just because it is general, take into account "the (varying) content of knowledge (relation to its specific object)". "But since truth concerns just this very content, it is quite impossible, and indeed absurd, to ask for a general test of such content" — for a criterion which is both sufficient and at the same time general is not possible. If sufficient, the criterion must be specific to each content; if general, it must hold good for all knowledge. Such a criterion, Kant concludes, would be self-contradictory.[11] How can there be, in that case, a logic of truth, and in what sense?

Kant goes on to assert that the purely logical criterion of truth (i.e., the agreement of knowledge with the formal laws of understanding) is the negative condition of all truth. No knowledge is true if it is not in agreement with these formal laws, e.g., if it is self-contradictory. But "further than this logic cannot go."[12] Logic teaches us nothing whatsoever, he writes a little later,[13] regarding the content of knowledge.

Kant's thesis at this point may be taken to mean that logic, whether formal or transcendental, cannot advance a theory of truth and that he himself does not give any. What he gives, in the transcendental Analytic, is a theory of objective validity. An objectively valid judgment brought about by application of the categories under appropriate conditions of sensory perception, may be either true or false. Transcendental Analytic does not guarantee the truth of a judgment, it assures its objectivity, i.e., the property of being either true or false. We have, for example, no theory of *truth* of our contingent, empirical judgments.

Or, one may regard Kant's point to be that while transcendental logic is a logic of truth (as he explicitly says with regard to the Analytic part of Transcendental Logic), formal logic has nothing to do with truth. This seems to be the interpretation proposed by G. Prauss in a paper which first drew attention to the entire issue.[14]

Alternately, and this is the interpretation suggested by Hans Wagner in his criticism of Prauss's paper,[15] one may take Kant to mean that while formal logic gives a general but not sufficient criterion of truth, transcendental logic supplies what formal logic fails to deliver, i.e., the content or matter of knowledge. Transcendental logic is the logic of truth, according to Wagner, in the sense that without the transcendental-logical there would be no valid relatedness to the object, and therefore, no possibility of truth of any specific empirical cognition. But transcendental logic cannot of course give a sufficient criterion of truth.

This discussion of the possibility of a logic of truth within the context of Kantian transcendental philosophy lies very close to the problem Husserl confronts. Husserl will have a logic of truth *within* the scope of formal logic, and then go on to develop a transcendental logical account of the possibility of formal logic. Is a purely formal logic of truth at all possible? And if it is, what would it look like?

III

Generally speaking, Husserl's logic of truth may mean any of several things: (i) in the first place, it may mean a theory of evidence (since Husserl does indeed regard the true and the evident to be in some sense identical); (ii) or it may have to include, even if it may not be the same as, a theory about the material or contentual (*inhaltlich*) conditions of a proposition's being true (the purely formal conditions having been given by the logic of consequence); (iii) or, a formal logic of truth may be none other than, and cannot be more than, a model of the pure logic of non-contradiction, in which the variables of the system are assigned objects (individuals and properties) in the world, so that to each model there corresponds one possible world. In this case, the logic of truth would be none other than a possible world semantics. Finally, (iv) Husserl's logic of truth is, and should be, none other than a phenomenological theory of cognition in its striving after truth, and therefore, identical with a phenomenological theory of knowledge.

A preliminary look at these alternatives shows that (ii) and (iv) cannot yield a formal theory of some sort, and yet Husserl does regard his logic of truth to be a stratum within formal (apophantic) logic. A general (if not formal) theory of material conditions of truth is open to the Kantian sort of skepticism. If these material conditions, in Kant's sense, are the *a priori* intuitions of space and time and belong to transcendental aesthetics, Kant keeps them apart from, and outside of the purview of transcendental logic. Husserl, it would seem, would include something he calls "transcendental aesthetic" within his transcendental logic.[16] A theory of the material conditions of truth would fall within Husserl's transcendental logic. But such a material truth logic cannot be a mere extension or deepening of formal truth logic. It has to be a critique and foundation of the latter.[17]

If (iv) were what Husserl intended by his logic of truth, then the Sixth Logical Investigation would be as close an approximation to it as anything he has ever given. But the Sixth Investigation cannot be said to be a formal theory in any case.

So only the alternatives (i) and (iii) remain. Husserl has not worked out (i), but there are indications in the *Logical Investigations* of what it may look like. It would have to contain such theorems as:

(1) 'p is evident to S at time t' implies 'p is true'; but
(2) 'p is true' does not imply 'p is evident to S at any time t';
(3) 'p is not true' implies 'p cannot be evident to S at any time t';
(4) 'Both p and ~p cannot be evident to any S at any time t'.[18]

More recently, following Brentano, Chisholm has developed such a formal theory of evidence.[19]

The other remaining alternative, i.e., (iii) model theory, is much more promising, and has been as such worked out with far greater success than (i). Commenting on the role of truth in contemporary logic, Kripke writes:

> Once a formal language has been defined syntactically, the main *semantic* demand is that an *explicit definition (characterization) be given (in understood language) of which sentences are true.* (If relevant, other concepts such as satisfaction, truth-in-a-model, etc., may be defined also.)[20]

Following this remark, one could surmise that even if Husserl did not explicitly make the distinction between syntax and semantics, his distinction between logic of consequence and logic of truth may be construed as capturing part of that distinction. Within a system of formal logic, validity may be defined syntactically as *derivability* (from the axioms of the systems and in accordance with the rules of inference of the system) and semantically as *true* in all interpretations of the uninterpreted system. The former would yield the notion of theoremhood, the latter the notion of logical truth.[21] The idea of 'logical truth', explicated as 'being true in all possible worlds or models', entails the idea of 'truth' (empirical truths being propositions which are true in some models, false in others). Tarski correctly points out that 'provability in a formalized deductive system' cannot be the same as 'being true' in the ordinary sense of that expression, because in its ordinary usage a true sentence must conform to the principle of excluded middle, which is not valid in the domain of provable sentences.[22]

Now, in order to be able to decide whether Husserl's concept of truth logic is indeed any of these two (i.e., a formal theory of evidence or a model theory), we need to take into account only three things Husserl says at some length in connection with truth logic. These concern, first, the concept of evidence pertaining to this stratum of logic; secondly, the precise formulation of the fundamental principles in this stratum; and, finally, a distinction between syntax and semantics that Husserl draws (a distinction which may indeed be very different from that to be found in more recent logic).

IV

First, as to 'evidence': although this essay is not intended to go into the details regarding the three different sorts of evidence that Husserl distinguishes corresponding to the three strata of formal logic, it will perhaps help us to interpret his concept of truth logic if we can understand this theory of evidence. Put succinctly, the theory consists in the following correlations:

(a) the pure theory of forms of judgment....judging confusedly;
(b) the formal logic of non-contradiction...judging distinctly;
(c) the formal logic of truth...............judging clearly.

At level (a), any well-formed sentence expresses a proposition. Sentences which at level (b) are inadmissible as involving contradiction are admissible at level (a). The domain of (a) consists of all well-formed sentences, no matter if they are formally inconsistent (in which case they would not belong to (b)) or if they, though formally consistent, are materially non-sensical (in which case they would belong to (b) but would be ruled out of (c)). In this broadest sense, to entertain a proposition (by "following" what the other says or by oneself uttering the sentence which expresses it) is to have it "vaguely" as an intended unity of sense. But since one "explicitly" performs the act of judging by positing the subject and then predicating the predicate of it (step by step, depending upon the complexity of either term), the earlier judgment, if it shows internal contradiction, is canceled, it is not any longer a genuine unity of meaning. Those unities of meaning which survive this process are explicitly distinct ones. It is these that form the domain of formal logic of non-contradiction. But it is quite possible that one has 'explicitly' performed an act of judging and thereby the unity of meaning has been made distinct, but one may still not be 'seeing' or 'intuiting' or 'seizing upon' the intended state of affairs. In other words, although the unity of meaning has been distinctly grasped, the intended objective state of affairs has not been either intuitively grasped or intuitively anticipated. When this last is achieved, the judging has become 'clear'. Such clear judgments alone can be established to be either true or false. A judgment such as 'Virtue is green' is consistent as a unity of meaning, but cannot achieve intuitive clarity, so it cannot be either true or false. For our present purpose, it is relevant to ask: what does Husserl exactly mean by 'clarity' and 'distinctness'? In what sense of 'clarity' is a clear judgment, and not merely a distinct one, a possible subject of predication of either truth or falsity?

A compound judgment is distinct, in Husserl's sense, if it can be explicitly performed. By 'explicit performance' he means actually positing the subject term, and then predicating the predicate of that subject. If an originally received belief (i.e., a judgment) was entertained but not explicitly performed by me, then it is for me a vague judgment. For most non-mathematicians the propositions 'There are transfinite numbers' or 'All prime numbers other than 2 are odd' are entertained as vague judgments, passively received from others. Only the mathematicians, especially the number-theoreticians, can transform them to explicitly performed judgments. To be able to do so, one must understand what is meant by 'prime numbers' or by 'transfinite numbers', so that one can then explicitly affirm what is to be predicated of them. What was in that case a vague judgment in the first place, becomes a distinct one. Note that it is not necessary that the movement shall always be in this direction: from vagueness to distinctness. One may arrive at any of the above judgments by spontaneous thinking, i.e., by explicit performance of thinking. Others may hear from him and assert it without fully thinking it through. In that case, they have only a vague judgment.

Now for our present purpose, what is relevant is that in course of explicitly performing a judgment, one may discover a contradiction in what was vaguely entertained. In such a case, one rejects the judgment, even if the explicitly inconsistent judgment is identical with the one which had been earlier vaguely entertained.

It is this distinct judgment in whose texture the presence or absence of contradictions is explicitly noticed, that according to Husserl is the focus of the logic of non-contradiction. In other words, in order to determine if two judgments p and q are consistent, incompatible or related by the relation of implication ('q follows from p'), it is first necessary to explicitly perform the two, i.e., to constitute, produce, each in actual thinking. Such a performance may show that the two are mutually contradictory — the one predicating B of A and the other predicating non-B of A.

For the purpose of the first stratum of formal logic, i.e., for what Husserl calls the theory of pure forms, the p's and q's remain unanalyzed. In this sense, they are vague judgments. The theory develops the laws determining possible operations on them. The theory of syllogism as a theory of judgment-complexes belongs also to this theory of pure forms.[23] Theory of syllogism, however, is first-order predicate logic, and so analyzes the p's and q's to such forms as (x)Fx, Fa, (∃x)Fx. Now Husserl might say, even these explicated judgments are still vague unities of meaning. What more is needed in order to detect a contradiction in p.~p or a consistency between (∃x)Fx and (∃x)~Fx?

The question I am raising may be reformulated thus: why does Husserl claim that a logic of non-contradiction requires more than *such* explication? The very point of formal logic, one may argue, is that by inspecting the sheer form of a judgment, i.e., without explicitly performing the judgment in Husserl's sense, one can decide whether the judgment under consideration is consistent or not. To determine "whether judgment-members included in a whole judgment ... are *compatible with one another, or contradict one another*",[24] or "*to seek out systematically the eidetic laws that* govern just the analytic includedness and excludedness"[25], requires some way of perspicuously analyzing the form of a given 'material' judgment. If this is what is meant by 'explicit performance', then Husserl's thesis is likely to be uncontroversial. But if Husserl is asserting the thesis that in order to determine the above mentioned analytic features one must posit the subject term, then affirm or deny the predicate term of it (in steps, depending upon the complexity of the subject and/or of the predicate terms), one is certainly requiring too much of the formal logician. Perhaps one is requiring precisely that psychological process to be undertaken, which the formal logician's tools of formalism are meant to succeed in making avoidable. Husserl perhaps wants to say that for this level of formal logic, a judgment which was in the first level an unanalyzed whole be subjected to step-by-step analysis, which would make its internal structure distinct. To this rather uncontroversial thesis, he would add that such step-by-step analysis (which the formal logician in any case carries out purely *symbolically*) it *must, in principle*, be possible to correlate a step-by-step actual thought process of 'constituting', 'bringing into being', or 'bringing into originary evidence' the structured unity of thought expressed in the judgment. This, I think, is a perfectly defensible claim.

Let us now consider what Husserl means by 'evidence of clarity' which is said to belong to judgments in truth logic. Clarity, we are told, concerns not the components of a judgment, but the state of affairs which is sought to be cognized by means of the judgment.[26] A clear judgment, in this sense, is evident judgment. In order to understand Husserl's thesis in this connection I will ask two questions. First, if a clear judgment is an evident judgment, then a clear judgment must be true. This however seems to be too strong a consequence for his purpose. He needs a conception of judgment narrower than the merely distinct unity of meaning, and wider than the true judgment. He needs to make room for the possibility of either truth or falsity (which is what Kant had in mind when he spoke of the 'objective validity' of a judgment). Second, it should be the case that some judgments that are permitted within the logic of non-contradiction, i.e., which are distinct unities of meaning, are not permitted within the logic of truth. What are these, and on what grounds are these excluded? What do these grounds have to do with 'clarity of evidence'?

To the second question, it seems Husserl's answer is that such materially non-sensical propositions as 'Virtue is green' are not excluded by a logic of non-contradiction. Since such judgments cannot be made clear or evident, they are excluded only by a logic of truth. (In the first stratum of pure logical grammar, even such materially non-sensical judgments are allowed as vague unities of meaning.) Now, this account can be questioned in the following manner. If a judgment in the logic of non-contradiction is a judgment which can be, or is actually and explicitly performed, then it is questionable whether one can explicitly perform, i.e., think step-by-step the judgment 'Virtue is green'. If Husserl's thesis is that it can be, i.e., if the proposition expressed is a distinct unity of sense, that must be because he is understanding 'explicit performance' in a purely verbal sense. As a matter of fact, Husserl seems to be saying so in *Erste Philosophie* where the distinction between the two logics is stated in a much sharper manner than in *Formal and Transcendental Logic.* He says that the analytical distinct-making yields a merely symbolic and verbal judging.[27] Certainly in a strict sense, I cannot, in thought, constitute a unitary meaning out of the sentence 'Virtue is green'. I think Husserl's claim on this matter is due to his belief that no purely formal-logical account can be given of the sort of contentual incompatibility this alleged proposition exemplifies. While that is indeed true, it does not help any more to assign that very task to another stratum of formal logic, namely, the formal logic of truth.

In view of these difficulties, I would suggest that a better organization of the structure of formal logic would be to divide the first stratum, i.e., the pure logical grammar into two sub-strata: first, a purely syntactic level of distinguishing well-formed sentences from those that are not well-formed; and then a semantic level which lays down the semantic categories and rules determining their compatibility and incompatibility in order to avoid materially non-sensical sentences. Given this ground work, we have excluded both sentences that are not well-formed and those that are well-formed but semantically non-sensical. Now the logic of non-contradiction may be regarded as dealing with the analytic relations amongst the consistent unities of meaning expressed by such syntactically and semantically permissible sentences. The logic of truth then would be dealing with the *same* unities of meaning but now from the perspective of their possible truth and falsity.

To the first question, namely, how is clarity of evidence to be distinguished from actual truth, I will reply by insisting that Husserl is concerned with both possibilities: the possibility of clarity of evidence that p and the possibility of clarity of evidence that non-p.

What I am suggesting then is that the logic of non-contradiction and the logic of truth differ not in this that some judgments permitted into the former are barred from the latter, but in the totally different perspectives from which they consider the same body of propositions. A logic of non-contradiction treats them as *propositions*, as mere unities of meaning, as ideal, abstract entities. A logic of truth considers them as contents of cognitions, as *judgments* (affirmations or denials) with claims to truth, but with the possibility of turning out to be false. I should add, the domain of pure logical grammar, in that case, should be taken to be the domain of *sentences* which are syntactically as well as semantically well-formed. Thus, on the account I am proposing, the three strata would be:

a pure grammar of logical *language*,
a logic of non-contradiction of *propositions*, and
a logic of truth of *cognitions*.

V

Husserl held that the same logical principles — such as the principle of non-contradiction — receive quite different formulations in the upper two strata of logic.[28] In the logic of non-contradiction, they are to be formulated without using the concepts of truth and falsity. Thus, the principle of non-contradiction may be formulated thus:

Of two contradictory propositions p and non-p, not both can be given in distinct evidence; or, both cannot have ideal 'mathematical existence'.

Since it is not at once clear why both the propositions 'There are transfinite numbers' and 'There are no transfinite numbers' cannot be said to express distinct unities of meaning, one may restate the principle thus:

A proposition in which two contradictory propositions are conjoined is not possible as a distinct unity of meaning, i.e., as a proposition proper.[29]

In the logic of truth, the principle receives the familiar formulation: not both p and non-p can be true. If one is true, the other is false. Not both can be brought to the evidence of clarity.

The two formulations are said by Husserl to be *analogues*.[30]

In the above formulation of the principle in the logic of non-contradiction, the idea of 'mathematical existence' is left unclear. The idea of existence is connected, in Husserl's mind, with the idea of 'properly effectible'.[31] That unity of sense exists, which can be properly constituted in step-by-step thinking.

The account I have proposed regarding the proper domains of the three strata of logic removes one possible objection against the two formulations given above of the logical principles. The objection may be raised that in both formulations we should still be talking of the same p and the same q. In the one case, we are assured that they are unities of sense, in the other that they are true or false. In that case, their domains would coincide. (This objection is particularly serious if, as I have also suggested, the semantic considerations of compatibility and incompatibility are to be transferred to pure logical grammar.) If the two domains coincide, the talk of 'ideal existence' or its absence can easily be translated into that of true or false. If, that is to say, in logic of non-contradiction only p has ideal existence, and not non-p, then in logic of truth only p is true and not non-p. If there is this nice correlation between ideal existence and truth, then the idea of a logic of truth becomes vacuous. Two answers are possible to this objection. In the first place, what does not have ideal existence is neither p nor non-p, but p.~p. This leaves it open that p may be true and non-p false, or vice versa.[32] In the second place, while ideal existence or its absence is asserted of propositions (so that p and q are to be construed as propositional variables), in truth logic they stand for cognitive claims (whose contents are propositions). Thus, there is a non-coincidence of the two domains.

VI

One of the reasons for the unsatisfactoriness of Husserl's conception of a logic of truth is that he for one thing characterizes it as formal logic, but also assigns to it tasks which are proper only for a transcendental logic. His accounts of a strictly formal logic of truth are sparse, and do not go beyond formulations of a few logical principles. His account of what a logic of truth could be, as a part of his transcendental logic, are more interesting. Let us briefly attend to these.

Transcendental logic examines the presuppositions of formal logic. As far as truth and falsity are concerned, these presuppositions are: (i) that a judgment is identifiable and re-identifiable as the same judgment; and (ii) that a judgment is decidable as true or false.[33] With regard to (i), Husserl writes:

> Logic relates, not to what is given only in active evidence, but to the abiding formulations that have been primarily instituted in active evidence and *can be reactivated and identified again and again*; it relates to them as objectivities which are henceforth *at hand*, with which, taking hold of them again, one can operate in thinking, and which, as the same, one can further shape categorically into more and more new formulations. At each level they have their manner of evident identifiability; ... (FutL, p. 185)

Further:

> What I have said, I have said; *I can at any time become certain of the* identity of my judicial meanings or opinions, my convictions, *after a pause in my thinking activity*, and become certain of them, in insight, as an abiding and always available possession. (FutL, p. 186)

The formal logician does not worry about this problem, for he deals with variables such as p's and q's, and even when he posits constants such as A and B, F and G, they are other *letters*. The assumption is that the identifiability of these letters by their shapes is all that matters. But what replaces these letters (if they are variables) and what the letters (if they are constants) name, are 'material' judgments, 'material' objects, 'material' predicates, and the identifiability of these is the presupposition Husserl has in mind. There may be a voluntaristic decision "this is what I mean, from now on this is what I shall mean". But one takes for granted that when you return to it in course of a chain of thinking (or proof, derivation, etc.) it must actually be the same. (Compare the question of identifiability and re-identifiability of perceptual objects.)

Connected with these is the further presupposition that "I can always return to it, do it again and again", that a proof can always be gone through (by adding 'and so forth'), e.g., "Given a cardinal number a, one can always form $a + 1$". Furthermore, when one says, as when the so-called principle of identity holds, that if A is true, it is true once and for all; or when one should be able to say, when I hold that a certain proposition is true, that is not my possession but everyone can entertain it — one makes claims to validity for all times ('once for all') of for all possible subjects ('for every one'), presupposing thereby that "everyone is in perfect harmony with everyone else". Husserl calls all these 'idealizing presuppositions', for as a matter fact, one just cannot carry out these processes always and endlessly.

When Husserl speaks of idealizing presuppositions, he should not be misunderstood as suggesting that the validity-claims of formal logic are just not to be trusted. What he really wants to bring to light are rather the presuppositions of those validity-claims. By bringing these to light, formal logic is made secure and well-founded. We discover what sorts of idealizing acts of consciousness make formal logic possible.

As regards 'decidability', it is also a presupposition of formal logic of truth that every judgment is decidable, i.e., can be decided as being either true or false. In the subjective locution which Husserl employs, it is capable of being brought to either positive or negative evidence. On this matter, he writes:

> We all know very well how few judgments anyone can in fact legitimate intuitively, even with the best efforts; and yet it is supposed to be a matter of a priori insight that there can be no non-evident judgments that do not "in themselves" admit of being made evident in either a positive or negative sense.[34]

Did Husserl himself accept the law of excluded middle? There is no evidence that he did, nor is there clear evidence that he did not. He discusses it as a presupposition of the standard formal logic of truth. He perhaps held it as belonging to the very sense of a formal apophantic logic. He saw his own task as a transcendental logician to question it (not with regard to its validity, but with regard to its sense), and this he did — first, by showing, in the text just quoted, what the claim to unconditional validity presupposes, and secondly, I should add, by limiting that claim. Not every judgmental unity of sense (which is syntactically permissible) is either true or false ('Virtue is green' is not). A judgment which violates the laws determining the compati--bility/incompatibility of *contents* (i.e., the semantic laws of permissible combinations of sense) is in violation of the law of excluded middle.[35] If that is so, then the applicability of the law of excluded middle presupposes that there are such compatibility/incompatibility, semantic, relations amongst content-senses, but such relations are grounded not in formal logic but in the way the experienced world is, as "an antecedent unquestioned matter-of-fact". What this *critique* reveals is that logic, without ceasing to be a priori, is still world-related — even if logic, as formal ontology, "must not assert facts of any sort, nor any *de facto* world, in its theories."[36]

Various remarks of Husserl as well as his thesis regarding "definite manifold" (which is defined in terms of 'decidability') suggest, when taken out of the context of his overall project, that Husserl — in clear conflict with the *Gödelsatz* of which he even in the late thirties does not evince any awareness — accepted the validity of the principle of excluded middle and of the decidability claim as well as of a purely axiomatic understanding of mathematics. However, this is not quite so. I would here venture the suggestion that Husserl's philosophy of logic and mathematics clearly passed through three stages, the first two of which, but not the third, are clearly noted in the literature. Husserl began (i) with an operational, computational, conception of mathematics, (ii) then rejected it in favor of an axiomatic understanding of both mathematics and logic in the middle nineties, but (iii) subsequently — while retaining that axiomatic interpretation with regard to *formal* logic, developed an inutitionist-constructionist theory of *transcendental* logic. What he did not do was to make his concept of *formal* logic intuitionist-constructivist, but *critiqued* formal logic (as he understood it) from the stance of a constructivist transcendental logic.

VII

Husserl's logic of truth (which is explicitly intended to be a formal logic of truth) should not be construed as but another name for his formal ontology.[37] This confusion is ready at hand. It is only natural to suppose that a discipline which lays down the structure of objects in general must have something to do with truth in its formal aspects. But that impression is deceptive. Formal ontology, as Husserl understood it, is a correlate of formal apophantic logic, both of logical grammar and of the logic of non-contradiction. It is a correlate in the sense that you need only a *change of attitude* to *transform* the apophantic logic to a formal ontology. The relation may be schematically represented thus:

apophantic logic ———-> formal ontology
1. logical grammar ———-> formal objective categories
2. logic of non-contradiction -> theory of objective manifolds
or theory-forms

Truth/falsity are not themes of formal ontology. The relation of formal logic (including truth logic) to transcendental logic may be represented thus:

{ 1. pure logical grammar
2. logic of non-contradiction } ←→ Formal Ontology

3. logic of truth *Ontology*
↓
4. synthetic 'material' logic
↓
5. transcendental aesthetics (a part of transcendental logic, contrary to Kant's plan)

In order to understand how truth logic, in the long run, leads to a transcendental aesthetics, i.e., a theory of the pre-predicative perceptual experience of the world, Husserl's argument may be briefly stated here.

(i) First, he proposes[38] a transitional link between the logic of non-contradiction and the logic of truth. The link consists in showing that every judgment is about something, and that this object-about-which (the replacements for 'S' and 'P' in 'S is P') is either itself a syntactically formed entity (e.g., 'The present President of the U.S.A.') or a core stuff free from any syntactic form (e.g., simply 'red', although that too may be taken to have syntactic (nominal or adjectival) form). The ultimate core entities are what Husserl calls *individuals*. All truth, in the long run, relates to them. Although formal mathematics has no interest in individuals (in the above sense), formal-logical theory of truth cannot but be interested in those individuals quâ individuals. To make these individuals intuitively evident, one needs to take recourse to *material* evidence (and not merely analytical evidence which makes unities of sense distinct) which is none other than *experience* in the strict sense.

(ii) One accordingly reduces truths of higher levels (of predicative judgments) to truths that relate directly to individuals.[39] (Compare Russell's principle of reducibility to acquaintance.) But the sense of this reduction must be precisely understood so that one does not violate what Husserl elsewhere calls the Principle of all Principles.[40]

(iii) Thus *ultimately* formal logic must be applicable to the domain of individuals. Such applicability belongs to the very sense of formal logic. In this sense, formal logic is world-related.

(iv) Prior to all judgment, there is the "universal experiential basis", "a harmonious unity of possible experience",[41] in which "everything has to do materially with everything else". It is this unity of experience which provides the ground of why some contents (content-senses) are compatible, some not. Thus concludes Husserl:

> *Formal-logical considerations* and theory, with their focusing on what is Objective, have nothing to say about that; but every one of their logical forms, with their S's and P's, with all the literal symbols occurring in the unity of a formal nexus, *tacitly presuppose* that, in this nexus, S, P, and so forth, have "*something to do with each other*" materially. (FutL, p. 219)

Thus truth logic rests upon a transcendental aesthetic. This is Husserl's response to the Kantian problem about whether a logic of truth is possible.

NOTES

[1] For a review of literature, see Heffernan, G., *Isagoge in die phänomenologische Apophantik*, Kluwer Academic Publishers, Dordrecht, 1989.

[2] Husserl, E., *Formal and Transcendental Logic*, e. tr. D. Cairns, Nijohff, The Hague, 1969. §20.

[3] Ibid., §16.

[4] Ibid., §14.

[5] Ibid., §22.

[6] Heffernan, 1989.

[7] Sokolowski, R.: 1974, *Husserlian Meditations*, Northwestern University Press, Evanston. Chs. 8-9.

[8] Kant, I.: 1787, *Critique of Pure Reason*, e. tr. N.K. Smith. B87 = A62.

[9] Ibid., B79 = A55.

[10] Ibid., A55/6 = B80.

[11] Ibid., A58/9 = B83.

[12] Ibid., A59/60 = B84.

[13] Ibid., A61 = B86.

[14] Prauss, G. (ed.): 1973, *Kant zur Deutung seiner Theorie von Erkennen und Handeln*, O.J., Köln. pp. 73-89 (*Kant Studien*, 60, 1969, pp. 166-182).

[15] Wagner, H.: 1977, 'Zur Kants Auffassung bezüglich des Verhältnisses zwischen Formal- und Transzendentallogik', *Kant Studien* 68, pp. 71-76.

[16] Husserl, 1929, pp. 291-2.

[17] Bernet, R., I. Kern and E. Marbach: 1989. *Edmund Husserl. Darstellung seines Denkens*, Felix Meiner Verlag, Hamburg. p. 49.

[18] Husserl, E.: 1900-1, *Logical Investigations*, e. tr. J.N. Findlay, Humanities Press, New York, 1970. Vol. 1, ch. 8, §51.

[19] Chisholm, R.: 1966, *Theory of Knowledge*, Prentice-Hall, Englewood Cliffs, N.J. Chs. 2 & 3.

[20] Evans, G. and J. McDowell (eds.): 1976, *Truth and Meaning: Essays in Semantics*, Clarendon Press, Oxford. p. 609.

[21] Haack, S.: 1978, *Philosophy of Logics*, Cambridge University Press, Cambridge. pp. 13-14.

[22] Tarski, A.: 1956, *Logic, Semantics, Metamathematics*, Clarendon Press, Oxford. p. 186.

[23] Husserl, 1929, §13a.

[24] Ibid., §14.

[25] Ibid., §14.

[26] Ibid., §13a.

[27] Husserl, E.: 1956, *Erste Philosophie. Erster Teil, Husserliana* VII, ed. R. Boeh, M. Nijhoff, The Hague. p. 25.

[28] Husserl, 1929, §20.

[29] Ibid., §20, fn.

[30] Ibid., §20.

[31] Ibid., Appendix III, §2.

[32] Ibid., §48.

[33] Ibid., §78; Sokolowski, 1974, p. 221f.

[34] Husserl, 1929, §77.

[35] Ibid., §90.

[36] Ibid., §92a.

[37] Contrast Bernet, R.: 1981, 'Logik und Phänomenologie in Husserls Lehre von der Wahrheit', *Tidschrift voor Filosofie* 43, pp. 35-89, esp. p. 57.

[38] Husserl, 1929, §82.

[39] Ibid., §83.

[40] Husserl, E.: 1913, *Ideas: General Introduction to Pure Phenomenology*, Book I, e. tr. Boyce Gibson, Collier Books, New York, 1962. §24.

[41] Husserl, 1929, §89b.

12. KANT ON "TRUTH"

(For Pranab Sen)

Kant's remarks on "truth" are sparse, and have given rise to questions such as whether he advocated a correspondence or a coherence theory of truth. There is one paragraph, however, in the introductory sections of the part on "Transcendental Logic" in the *Critique of Pure Reason*, which has given rise to conflicting interpretations. Some of the issues this paragraph raises concern the relation between formal and transcendental logic and the place of "truth" in each. I will begin by focusing upon this text.

After introducing the idea of transcendental logic as distinguished from "General logic", Kant proceeds to divide General logic into Analytic and Dialectic. At this point, in A58 = B82, he says that the question which "drove logicians into a corner", forcing them either to take recourse to "a pitiful sophism" or to confess their ignorance and therefore "the emptiness of their whole art" is "what is truth?" Then Kant goes on:

> The nominal definition of truth, that it is the agreement of knowledge with its object, is [here] assumed as granted; the question asked is as to what is the general and sure criterion of the truth of any and every knowledge.[1]

He proceeds to interject a remark about the importance of knowing what questions may reasonably be asked:

> For if a question is absurd in itself and calls for an answer where none is required, it not only brings shame on the propounder of the question, but may betray an incautious listener into absurd answers, thus presenting, as the ancients said, the ludicrous spectacle of one man milking a he-goat and the other holding a sieve underneath.

What is this absurd question, in our present context, which Kant has in mind? The question Kant has in mind, the one he regards "quite impossible, and indeed absurd to ask" is: What is the general test of truth — i.e., "a sufficient and at the same time general criterion of truth"? The reason for regarding this question to be absurd is clearly given by Kant thus:

> ... a general criterion of truth must be such as would be valid in each and every instance of knowledge, however their objects may vary. It is obvious however that such a criterion [being general] cannot take account of the [varying] content of knowledge (relation to its [specific] object). But ... truth concerns just this very content ... . (A58-9 = B83)

Kant concludes: "we must be prepared to recognize that of the truth of knowledge, so far as its matter is concerned, no general criterion can be demanded." (A59 = B83)

In this remarkable paragraph, there are three issues (amongst others) that stand out:

First: when Kant writes that the nominal definition of truth (rather, its *Namenerklärung*, is "here assumed as granted", what does "here" mean? Is it assumed and granted by Kant himself? Is it assumed by those who ask the "absurd" question regarding "the general and sure criterion of the truth of any and every knowledge"? Or, is it assumed by the logician (meaning by it, the formal logician)?

Second: if the formal logician cannot answer this question, is it Kant's view that transcendental logic can? But if the very question is impossible and absurd, how can even transcendental logic offer an answer? If Kant thinks it can, does Kant's transcendental logic as a matter of fact succeed in giving one?

Third: what is Kant's own substantive conception of truth? To these questions then we shall turn.

I

Several readers have so understood Kant's text on this point that by "here" (in: "Die Namenerklärung der Wahrheit, daß sie nämlich die Übereinstimmung der Erkenntnis mit ihrem Gegenstände sei, wird *hier* geschenkt und vorausgesetzt" (italicized by me)) Kant means the *Critique of Pure Reason* itself. On this reading (by Hegel, Brentano and Heidegger), Kant himself takes the word "truth" to mean "agreement of knowledge with its object." As opposed to this, Gerhard Prauss[2] argues (basing on the grammatical feature of the sentence) that Kant is not here presupposing his own understanding of "truth," but is rather referring to the handed-down historical situation in which the sceptic presents the question of truth for the logician in order "to corner" the latter into an uncomfortable position ("in die Enge zu treiben *vermeinte*"). Kant, according to Prauss, is not as such interested in that historical situation, but regards it as a *typical* problem: "when the sceptic asks the logician the question "what is truth?" — as it happened in antiquity — the word-meaning of "truth" as agreement of knowledge with its object is presupposed." On this reading, *this* question — which the sceptic asks the logician — Kant does *not* put for himself, i.e., for his tanscendental logic. If the very question put to the formal logician is "absurd" ("*ungereimt*"), then so are the answers to the question. Obviously then in his transcendental logic he will not be answering this question. Why then does he call his transcendental logic a logic of truth? Prauss, in answering this last question, suggests: it is not the question "what is truth?" that is *ungereimt*, but the question *as put to the formal logician* is so. The question about truth as formulated in transcendental logic, however, is perfectly in order and needs a meaningful answer. Note that if this is the case, then *the two questions are not the same*. "Truth" in any case concerns the content of knowledge, and the transcendental logic, as contrasted with formal logic, does *not* abstract — as Kant tells us — from all contents. Prauss shows where the two questions about "truth" — the one that arises for formal logic and which it cannot answer, and the one which transcendental logic asks and can answer — differ. The one, i.e., the question for formal logic, concerns (i) truth as opposed to falsity and (ii) each specific cognition separately (i.e., asks what makes *this* cognition true?). Transcendental logic, on the other hand, takes the question "what is truth?" to be the question "what makes a knowledge knowledge?", which again is the same as the question "what is it that makes a knowledge to be *either* true or false, i.e., objectively valid?" In this sense, "truth" is not opposed to "falsity". It means not agreement with its object, but rather its relatedness to an object. In this sense, truth is, according to Prauss, truth-difference, *Wahrheits-differenz* (an expression Kant certainly did not use). But Kant does use "*transcendentale Wahrheit*", which normally one takes to mean the truth of the *a priori Grundsätze*. Prauss takes it also to mean the possibility of either truth or falsity.

II

In order to be able to appreciate the controversial nature of Prauss's interpretation, let me briefly refer to a critique of Prauss's suggestion by Hans Wagner.[3] Wagner begins by pointing out that Prauss's interpretation implies a complete opposition between formal and transcendental logic, and if this were true then one would have difficulties in making sense of Kant's well-known procedure, in the metaphysical deduction, of using the table of formal logic as the "*Leitfaden*" for discovering the table of categories of transcendental logic. If the Kantian procedure has to be justified, then the Kantian conceptions of "knowledge" and "truth" must be closely related to the formal-logical conceptions.

Wagner rejects Prauss's contention that the *"Namenerklärung*" of "truth" is what is presented to the formal logician and not by Kant to himself. On the contrary, he interprets Kant's sentence to mean that Kant accepts this account of truth, and what Kant wants to determine is the condition of possible truth. Prauss, according to Wagner, is also mistaken in thinking that what leads to absurdity is the question regarding the nature or essence of truth; it is rather the question about a general *criterion* of truth that leads to the absurdity Kant points out. Consequently, Kant should not be read as if he wants to distance himself from formal logic, or as though he wants to take the question of truth entirely out of the domain of formal logic. Formal logic indeed, according to Kant, gives the merely negative condition of truth. To underestimate formal logic in this regard is, at the same time, to overestimate the role of transcendental logic, and both moves are mistaken. Transcendental logic cannot, any more than formal logic, supply a general and sufficient criterion of truth. Over and above the negative condition of truth that formal logic gives, transcendental logic supplies further rules which make possible valid relatedness to an object, and these rules concern the content of knowledge. But even these transcendental-logical conditions remain far from being the criteria of the actual truth of a cognition. They are necessary, but not sufficient conditions for truth. What is still needed is *intuition* but intuition cannot be a *general* criterion for truth, it is specific in each case. Transcendental logic is then logic of truth *only in the sense* that without the transcendental-logical no truth would be possible, but we still do not have a necessary and sufficient criterion of truth.

Note the difference between the interpretation by Prauss and that by Wagner: according to Prauss, if I understand his position correctly, formal logic is concerned with the truth (as opposed to falsity) of each specific cognition, whereas transcendental logic is concerned with "objective validity" (I prefer to use this perfectly Kantian term rather than "truth which is *not* opposed to falsity") of cognitions as such. Wagner understands Kant as holding that both formal logic and transcendental logic are concerned with the same "truth": formal logic gives negative conditions for a cognition to be true, transcendental logic *adds* the positive rules which the cognition (which has already satisfied the formal-logical negative conditions) must satisfy in order to have objective validity (i.e., relatedness to an object), but we still fall short of having a universally necessary and sufficient criterion which of course just cannot be given.

III

Before I decide between these two interpretations, the question still remains: does Kant himself accept the *Namenerklärung* of "truth" as consisting in the agreement of a cognition with its object? There is plenty of evidence to suggest that the *Namenerklärung* is not merely presented by the sceptic to the formal logician "in order to drive him to a corner", but is one which Kant also accepts. In the *Critique of Pure Reason*, there are at least two places where Kant seems to leave us in no doubt about this. At A191 = B236, he writes: "Since truth consists in the agreement of knowledge with its objects, it will at once be seen ... ." I need not quote the whole sentence. The antecedent of the hypothetical is not a counterfactual (i.e., Kant is not saying "if truth were ... .") but is asserted to be true. At A642 = B670, he writes "... while the categories lead to truth, that is, to the conformity of our concepts with the object ...", a remark which again confirms the ascription to Kant himself of the "*Namenerklärung*".[4]

Kant's concern with the problem of truth did undergo a historical development which I will briefly recall.[5] In the Blomberg Logic Lectures[6], Kant raises the sceptical objection to the idea of correspondence that knowledge cannot be compared to an object "outside us", but only with the known object. In the Logic Philippi (which was given after the Blomberg Logic), this sceptical objection is stated as an "argument", and is regarded by Kant as helping us to recognize the true nature of truth.[7] The first attempt to "solve" the problem is also suggested in these lectures: "truth is the agreement of cognitions of objects amongst themselves. For what we call objects are only our cognitions." These lectures come to the conclusion: "Truth is the agreement of knowledge with the universal laws of Reason."[8]

Kant goes on to say that there are two steps for checking the universal truth (i.e., truth for everyone, not for me alone): first I should compare my knowledge of an object with the knowledge of other objects which have already been taken to be true; secondly, I would see if my cognition agrees with that of the others. "In general," he writes, "that is true which holds good of everyone. What has only private validity is only appearance."[9] Then he goes on to lay down the following rules: if all consequences of a cognition are true, then the cognition is true. If only one consequence of a cognition is false, then the cognition is false. When many consequences of a cognition are true, the cognition is probable. But since one cannot examine *all* the consequences, one can infer the truth of a cognition from the truth of consequences only by approximation.[10] Very soon afterwards, in the famous letter to Marcus Herz of February 21, 1772, Kant asks "whence comes the agreement which (the pure concepts of the understanding) should have with objects which are not brought into being through them ... ?"[11] The reply of course is the main thesis of the *Critique of Pure Reason*. But in between the letter to Herz and 1781, there are reflections which mark the progress of his thinking. Reflection 2124 runs thus: "Knowledge agrees with the object, if it agrees with itself."[12] Reflection 2128 says: "the objective criterion of truth is the agreement of representations in a judgment amongst themselves in accordance with universal laws of understanding and of reason."[13] Reflection 2155, dating from the years 1776-1778, moves closer towards the *Critique*: the agreement of knowledge with itself is the formal criterion of truth and is only "negative", the material criterion of truth is the agreement of judgment with intuitions, "also nicht formale Tautologie und Identität".[14] Reflection 2161 makes the point still clearer: "The material criterion of truth consists in the agreement of knowledge with the representations which directly relate to the object, and so in the agreement with intuitions and perceptions."[15] The question discussed at the beginning regarding the general and sufficient criterion of truth appears in the Pölitz Logic of 1789.[16]

In the *Logik Pölitz* (1789), Kant asks the question (raised in the *Critique of Pure Reason*): what is the criterion of truth? For the logician, he says, the material criterion would lie in the agreement of knowledge with its object, the formal criterion in the agreement of knowledge with itself. Is there a universal material criterion? That seems to be impossible. The agreement of knowledge with itself, i.e., with the general laws of understanding, is said here to be a formal criterion.[17]

This discussion clearly shows that while Kant held on to the correspondence notion of truth, he nevertheless gave what has been called a "coherence-theoretical grounding of the correspondence theory of truth."[18] Correspondence remains the *Namenerklärung;* coherence gives the material criterion of truth. However, when one says this, one should be clear that Kant's theory cannot be interpreted along the lines of the more well-known varieties of coherence theory.[19]

IV

For "a coherentist grounding of the correspondence theory of truth", the following texts in Kant are relevant:

In *Prolegomena* §13, Anm. III, Kant writes: "The difference however between truth and dream is constituted not through the property of representations to be related to objects, for this property is the same in both cases, but rather through the connection of representations in accordance with rules which determine the interconnectedness of representations in the concept of an object and in so far as they are able to function in an experience or not." This text shows that for Kant "relatedness to an object" is common to both truth and dream, to truth as well as to falsity, and so cannot be simply regarded as truth. What distinguishes truth is that the representations function in an experience in a rule governed manner. I think, the last phrase in the above text is important, for in a false cognition also an object is represented and in so far there is a connection of representations in the concept of an object. But only a true cognition can function in an experience, where by "an experience", Kant means "one temporally interconnected *durchgängig* conscious experience".

In A737 = B765, speaking about how at all any valid synthetic judgment can be established, Kant says that the principles underlying them can be established "only indirectly through relation of these concepts [= categories] to something altogether contingent, namely, *possible experience*. If therefore there is an ultimate criterion of material truth, that criterion must relate to the idea of *an experience*. No particular judgment can then be, as a particular judgment, true by itself. What in the long run is the ground of its truth is "the universal interconnectedness of experience".

At A104, after saying that we must make clear to ourselves what we mean by "an object of representations", and after pointing out that "this object must be thought only as something in general = x, since outside our knowledge we have nothing which we could set out over against this knowledge as corresponding to it", Kant continues:

> The object is viewed as that which prevents our cognitions from being haphazard or arbitrary, and which determines them *a priori* in some definite fashion. For in so far as they are to relate to an object, *they must necessarily agree with one another,* that is, *must possess that unity which* constitutes the concept of an object. (emphasis by me.)

It is important to remember, when ascribing to Kant a sort of coherence theory, that Kant does not believe that agreement with judgments of others is anything but a subjective criterion of truth. Thus in Reflections dating from the time of the *Critique of Pure Reason*, Kant writes, under Reflection 2128, "The subjective criterion of truth is the agreement of a judgment with others whether of the same subject or of other subjects." In Reflection 2148, he says that agreement with "fremder Vernunft giebt kein Kennzeichen (Beweis) sondern Praesumption der Wahrheit."[20] In Reflection 2176, agreement with others' judgments is said to be external and subjective[21], contrasted with the logical and "inner" criterion which is, in Reflection 2182, said to be three-fold: that a cognition does not contradict itself (for "judgments"), that it has a "ground" (for "propositions") and that it does not have false consequences (to which Reflection 2185 adds another, namely, that its contradictory has false consequences.[22] But these are still formal-logical criteria. What about criteria for material truth? As noted earlier, Reflection 2161 says that the material criterion of truth consists in the agreement of knowledge with representations which directly relate to the object — and therefore with intuitions and perceptions.[23] But intuitions and perceptions are different from case to case, and so do not provide us with a general material criterion.

There is one author, Stuhlman-Laeisz, who contends that the Law of Sufficient Reason (or, the *Satz von Grund*), which according to Kant provides the "logical interconnection" of judgments and positively determines the logical truth of a judgment[24], is also *sufficient* for the material truth of knowledge. Stuhlman-Laeisz appeals to the *Logik Busolt*, where Kant writes: "hingegen ist bei der Wahrheit ... die Gründlichkeit aber ein hinreichendes Zeichen derselben."[25] In this context, "Wahrheit" is taken to mean "material truth". Thus writes Stuhlman-Laeisz:

> Wenn ich einen Satz habe, von dem ich nicht weiss, ob er ('materiell') wahr oder falsch ist, zugleich aber andere Sätze kenne, von denen ich dies weiss, denn kann es mir mit Hilfe des 'Satz vom zureichenden Grundes' gelingen, über Wahrheit und Falshheit des fraglichen Satzes zu entscheiden.[26]

Such an interpretation which ascribes to Kant a formal logical criterion of material truth is obviously in conflict with the various statements of Kant that we have already taken into account.[27] The crucial question is, how did Kant understand the principle of sufficient reason?

In his reply to Eberhard and in his letter to Reinhold dated May 12, 1789, Kant distinguishes between two senses of the principle of sufficient reason: a logical sense and a real sense. In the logical sense, the principle is to the effect that every proposition must have a ground, which Kant regards as an analytical truth. In its *real* sense — which is also the material (and also the transcendental) sense — the principle is that "everything must have its reason" which is a synthetic proposition which itself needs justification. In the real sense, the principle is nothing other than the idea of an interconnected system of experience.[28]

We therefore have to connect Kant's idea of *material* truth with his idea of a possible experience. I have already referred to the texts of *Prol.*, Anhang 375 and *Critique of Pure Reason*, B765. Let me recall that for Kant there is only *one* "possible experience". In the *Handschriftliche Nachlass* from the years 1796 - 1803, Kant notes: "Nicht Erfahrungen in Plurali sondern Erfahrung." Also: "Nur ein Gott, eine Welt, eine Erfahrung."[29] A little later, more pointedly: "[die] mögliche Erfahrung nur eine ist."[30] In the light of these, then, we can understand the text in the *Critique of Pure Reason*:

> There is one single experience in which all perceptions are represented as in thoroughgoing and orderly connection, just as there is only one space and one time in which all modes of appearance and all relation of being or not being occur. (A110)

This one experience is then said to be a "thoroughgoing synthetic unity of perceptions" (A110); and "the *a priori* conditions of a possible experience in general are at the same time conditions of the possibility of objects of experience." (A111) Furthermore, "that an object be given" — understood as signifying immediate presentation in intuition — "means simply that the representation through which the object is thought relates to actual or possible experience". (A156 = B195)

Thus a synthetic judgment is materially true only in so far as it conforms to the conditions of possible experience, i.e., belongs to the interconnected structure of experience. There are then two aspects of Kant's "coherentist reconstruction" of the correspondence notion of truth: for one thing, various representations must be synthesized under the concept of an object, and these representations must, in the long run, relate to an intuition; for another, the knowledge of the object must fit harmoniously into the contexture of one possible experience. *Both* are needed for "material" truth. In the case of dreams (and hallucinations), we have the former and not the latter.

V

Let me briefly return to the idea of "transcendental truth". One of the merits of Prauss's paper is that it emphasized that *"objektive Gültigkeit"*, according to Kant, is not truth. Even a false judgment "It is now snowing" (when it is not) is objectively valid, it is a judgment in which the categorial forms are employed. As judgments, irrespective of their truth and falsity, "It is snowing" and "It is not snowing" are both "objectively valid". What transcendental logic does then is to give the conditions for the possibility of truth and falsity, i.e., for the feature of judgments to be *either* truth *or* false, i.e., for their "objective validity". This is what Prauss takes to be "transcendental truth" which is neither *a priori* truth nor empirical truth. Thus, if a judgment P is in accord with the fundamental principles (*Grundgesetze*), then it is either true or false, if not, it is neither. This fact that Kant's "objective *Gültigkeit*" is not truth (as opposed to falsity) but the property of being either true or false has been recognized by many (e.g., Emil Lask). The Kantian principles being what they are, nature could have been very different from what it is. In other words, what Kant demonstrates, e.g., is the validity of the causal principle, but it is perhaps consistent with the *a priori* validity of the *Grundgesetze* that the specific causal laws be different from what they in fact are. That is to say, what Kant demonstrates is the objective validity of the causal principle, but not of the specific causal laws (such as "bread nourishes" or "fire warms") which obtain in nature as nature happens to be.

There are two problems for such an account, both of which have been pointed out by Stuhlmann-Laeisz. On this account, a judgment which contradicts the principles of the transcendental Analytic is neither true nor false (for the transcendental Analytic gives the necessary and sufficient conditions of being *either* true *or* false.) However, Kant has examples of such judgments (i.e., of judgments which do not satisfy the principles of the transcendental Analytic) which he also regards as false. The judgments "The world is finite" and "The world is infinite" are both taken by him to be false (A503f = B531f). Secondly, the fundamental principles of the transcendental Analytic are regarded by Kant as being *a priori* true (A157 = B196f), and their negations are false. Each of these negations, since it contradicts a *Grundsatz* could not be, however, on Prauss's interpretation, either true or false.

To the first objection, I would reply that the sense in which, on Kant's view, the two judgments "The world is finite" and "The world is infinite" are both false is different from the sense in which the judgment "Snow is black" is false. The same could be said of a judgment which negates a *Grundsatz*, e.g., the judgment "There is an event which is *not* preceded by another event in accordance with a rule". The latter is false, but again, in a sense that is different from the sense in which "fire cools" is false. In order to bring out this sense of falsity (let us call it falsity$_2$ as distinguished from the ordinary sense of falsity to be called falsity$_1$), let us recall what Kant tells us about falsity$_1$.

A false$_1$ judgment, according to Kant, does not agree with "its own object" to which it is related, although it "contains something which could very well be true of other objects." (A58 = B53) Kant seems to have held two theses regarding false$_1$ judgments: that a cognition cannot be entirely false, so that a false judgment must be only in part false and must contain a degree of truth; and that a false cognition has also an object.[31] I will not, for my present purpose, discuss all the philosophical problems that arise out of these two theses. The reason I want to limit these claims to falsity$_1$ is that in the case of false$_2$ judgments like "The world is finite", *there is no object* referred by "the world". It, as well as the contradictory "The world is infinite", are *totally* false. Likewise, "an event which is not preceded by another event in accordance with a rule" is not — if Kant's argument is correct — a possible object.

I will therefore support Prauss's reading on this matter, and say that the transcendental Analytic gives the necessary and sufficient conditions of being *either* true$_1$ or false$_1$. Judgments which do not satisfy them are false$_2$. The difference between a false$_1$ judgment and a false$_2$ judgment is that the former has its own object (of which it is not true) whereas the latter has no object of its own. A judgment then may be false because what it predicates of its object does not belong to it, or because there is *no object* at all of which the predicate is allegedly being asserted. "Fire cools" belongs to the first type, "The world is infinite" to the latter. The Principle of Excluded Middle holds good of the former, not of the latter ("The world is not infinite" being equally false$_2$). (There is a minor worry likely to arise from the second type: does "The unicorn is carnivorous" belong to it [since there is no object at all of which the predicate is being asserted]? "The unicorn", however, is a Kantian object, even though there are *really* no unicorns.[32] The judgment, therefore, is objectively valid, but false$_1$.)

VI

In this section, I want to reflect on the idea of "possible experience" as a criterion of truth$_1$. Can one say that a judgment that is "objectively valid" is

$true_1$ when it conforms to "possible experience" and $false_1$ when it does not? As we noted, in a text cited earlier in this paper Kant regards relatedness to an object as being common to both $true_1$ and $false_1$ cognitions. What distinguishes $true_1$ cognitions from $false_1$ cognitions is that the former functions in *an* experience in a rule governed manner. By "*an* experience", Kant means "one temporally interconnected *durchgängig* conscious experience". This "universal interconnectedness of experience" or "possible experience" is the ultimate ground of $truth_1$ (as distinguished from $falsity_1$).

To understand Kant's point here we must understand exactly what Kant means by "possible experience". First of all, Kant does *not* mean that the truth of a judgment is to be measured by agreement with the judgments of others. Kant regards this as a subjective criterion. Nor does he mean that a judgment is true if it agrees with its object — for that would imply a realism about objects which Kant cannot presuppose in his theory. Thus the talk of "possible experience" situates us midway between the subjective opinions of cognizing subjects and the alleged object-in-itself. Positively, we can, following Kant, say the following things about "possible experience":

In the *Prolegomena* §5, Kant describes experience as "a continuous synthesis of perceptions," not "mere empirical putting together (*Zusammensetzung*) of perceptions", but rather as perceptions connected together in accordance with the "fundamental principles" such as the Analogies of experience. But an absolute totality of all possible experience is never itself an experience, but rather a necessary problem for reason.[33] *There is however only one "possible experience"*. When one speaks of many experiences, they are only so many perceptions which belong to one synthetic continuous unity.[34] Furthermore, the *a priori* conditions of possible experience are also conditions of the possibility of objects of experience.[35]

If we put these characterizations together, then we can say that a true cognition belongs to the universal interconnectedness of one possible experience, while a false cognition (while still possessing objective reference, and while being logically consistent) is rejected by, i.e., cannot fit into, this interconnectedness governed by the fundamental principles such as the *Analogies*. But it is also a coherence which, to my mind, is compatible with fallibilism[36] with regard to empirical truths, for since the idea of a completed whole of experience is inadmissible on this theory, the ongoing process of experience leaves open the possibility of revising a truth-claim already accepted. We thus have a coherentist interpretation of the idea of correspondence (which still can be retained as a *Namenerklärung*.)

It is time to deal with a possible objection to the above. It may be said that what the idea of possible experience accounts for is the *possibility* of a cognition's being true$_1$, but whether a cognition is actually true$_1$ depends not upon whether the cognition satisfies the condition of possible experience, but upon whether it agrees with its actual object, i.e., if it has as its content an actual perception of whatever is its object. The fundamental principles then lay down the necessary conditions of a cognition's being true$_1$, but those conditions are not sufficient: one needs to take into account an actual intuition of the appropriate object. In other words, granted that no cognition could be materially true$_1$ (over and above being logically consistent) unless it also satisfied the conditions of possible experience; yet a cognition which did so might still fall short of being actually true$_1$.

This objection clearly is not valid as against the claim — which Kant explicitly makes (A156 = B195) and several times reiterates — that "The *possibility of experience* is, [then], what gives objective reality to all our *a priori* modes of knowledge". In the case of these "*a priori* modes of knowledge", their possibility and actuality coincide. After demonstrating that these *a priori* modes of knowledge are also the conditions of the possibility of experience, one does not need to show anything else in order to prove their truth (= "objective reality"). But it is otherwise with empirical judgments: in their case, it would seem, demonstration of the possibility of their being true$_1$ is not equivalent to demonstrating their truth$_1$.

There are two possible replies to this objection. For one thing, one may reply by pointing out that if a cognition belongs to the structure of possible experience, it is not, for that reason, merely *possibly* true, it is not the case that it is merely possible that the cognition is true. On the contrary, what is being maintained is that as belonging to the structure of possible experience, it is actually true$_1$. If this is so, then we need not look for any more conditions to be satisfied. Further — and this would be the second response — the objection makes a distinction between possible truth$_1$ and actual truth$_1$ (in general between a possible ϕ and an actual ϕ) that is not permissible within the Kantian framework. In the Kantian modal theory, the range of possibility and the range of actuality should coincide. Or, better still, what is possible in the strictly Kantian sense (not in the standard sense of logical possibility) is also actual.[37] In that case the objection relies upon modal concepts which are inadmissible in the Kantian context.

I think it is extremely hard to precisely assess the value of these two responses. But I will try to comment on the first one. Consider an empirical cognition c_i occurring at time t_i. Let us assume that the cognition c_i is not only "objectively valid" in the sense of Prauss but is also $true_l$. We are trying to understand what would be a Kantian explication of the sense of $truth_l$ (of c_i) (besides the standard *Namenerklärung*). The explication suggested by the coherentist reading of Kant amounts to this:

> c_i is $true_l$ iff c_i belongs to the *one* interconnected structure known as "possible experience".

Let us try to understand this explication a little better. Since the interconnectedness structure is open-ended, one can only say:

> c_i is $true_l$ iff c_i belongs to the interconnected structure of "possible experience" at time t_i (or, better still, up to t_i).

Beyond t_l, all that we have are the anticipated *a priori* yet-to-occur experiences (and objects) in accordance with the fundamental principles. For an empirical cognition to be $true_l$, then, is nothing but its belonging to the interconnected structure of experience *up to the temporal location of that cognition*. If that is so, the first response to the objection is right on target. A $true_l$ cognition belongs to, and a $false_l$ cognition is rejected by the interconnected structure of *one* possible experience. What is wrong with the objection is that it takes each single empirical cognition to be able to contain within it the condition necessary for its being $true_l$, the condition namely that it be an intuition of its own object. But this requirement would seem to be un-Kantian in view of the fact that a cognition's being of the object it cognizes is the result of a whole gamut of synthetic operations — operations not alone of welding intuitions together, but also of tying objects to other possible objects — as well as their perceptions — into one experience. Recall Kant's statement, earlier quoted, at A737 = B765 that no particular judgment can be, as a particular judgment, true by itself. Thus the coherentist reading stands.

VII

One reason why a coherentist interpretation of Kant's theory of $truth_l$ of empirical propositions is not generally put forward is that Kant is concerned, throughout the Transcendental Analytic, with the *objective validity* (i.e., objective applicability) of the pure concepts of the understanding (and of the fundamental principles) as well as the objective validity (and $truth_l$) of *all* cognitions (*including empirical cognitions*). Since these three sorts of tasks are not clearly separated in the exposition, and the problem of empirical $truth_1$ generally remains, textually, in the background, Kant appears not to have offered a general theory for it. The appeal to "possible experience" is then regarded as a part of the demonstration (or deduction) of the objective validity of the categories and of the fundamental principles — indeed, of all *a priori* cognitions, and consequently the concept of "possible experience" is taken to have no direct bearing upon the problem of $truth_1$ of empirical cognitions. My thesis in this paper is that this last claim is just mistaken. The concept of "possible experience" provides an explication — in non-formal terms — of the $truth_1$ of empirical cognitions as well.

How does my interpretation bear upon the controversy between Prauss and Wagner? Prauss, as said before, holds that while formal logic gives the (negative) conditions of truth (as opposed to falsity), transcendental logic is concerned with the conditions of objective validity (i.e., of the property of being either true or false, of what he calls *Wahrheits-differenz*), not of each specific cognition (which Kant, in A58-9 = B83, said it was impossible to give) but of cognitions as such. Wagner, on the contrary, understands Kant to be giving, first — along with formal logic — negative conditions of truth, and then — in transcendental logic — positive conditions for truth. But even the latter falls short of providing sufficient conditions thereof — which must include actual empirical intuitions.

I agree with Prauss that "objective validity" is not to be identified with "truth". What transcendental logic and its machinery of *a priori* concepts and principles give are necessary and sufficient conditions of a cognition's being either true or false, i.e., of being an "objective representation", of being related to an object (which characterizes both true and false judgments). However, Wagner is right in claiming that the transcendental logic is a logic of truth in the further sense that it does give a necessary condition of $truth_1$ in terms of "possible experience". That this necessary condition is also sufficient is the further thesis I ascribe to the Transcendental Analytic.

There remains, at the end, an anxiety regarding my thesis that needs to be set to rest. This anxiety rests on the distinction, noted earlier in this essay, between possible truth and actual truth. If this distinction has any point, it is that while the Kantian Transcendental Analytic lays down the conditions necessary for a cognition to be either true or false (i.e., to be possibly true, or also possibly false), we still need, in addition, conditions for it's actually being true, and the latter must include actually intuiting the object of the cognition. But, the argument would continue, conditions necessary for actual intuition of the object cannot be laid down by any logic — formal or transcendental. Now, "possible experience", as understood by Kant, can be both necessary and sufficient for truth$_1$ of empirical cognitions, if it includes not merely the possibility of givenness, but also actual givenness. My claim is that it so does. "Possible experience" designates that progression of experiences (and of objects of those experiences) which, starting with the actually now given proceeds in accordance with the Analogies of Experience both backward and forward, i.e., into the past and the future. It is as thus construed — and not understood as a conceptual system of abstract possibilities — that "possible experience" is all that one needs to serve as the contexture which would serve to demarcate cognitions that are true$_1$ from those that are false$_1$. It is not therefore true to say that "possible experience" is the principle of the mere possibility of truth$_1$; it, as a matter of fact, determines actual truth$_1$.

The pre-critical idea that a cognition is true if it corresponds to its object, is then, not negated but *aufgehoben* (and so preserved) in the critical theory that a cognition is true if it agrees with "possible experience".

NOTES

[1] References to the *Critique of Pure Reason* are inserted within the text in terms of A and B edition paginations. English translations generally follow Kemp Smith's. My changes are inserted within brackets. Note that some expressions within brackets are Kemp Smith's own.

[2] G. Prauss, "Zum Wahrheitsproblem bei Kant", *Kant-Studien* 60, 1969, pp. 166-182. (Reprinted in: Prauss (ed.), *Kant zur Deutung seiner Theorie von Erkennen und Handeln*. Köln O.J., 1973, 73-89.)

[3] H. Wagner, "Zu Kants Auffassung bezüglich des Verhältnisses zwischen Formal- und Transzendentallogik". Kr.d.r.V. A57 - 64/B82 - 88, *Kant-Studien* 68, 1977, pp. 71-76.

[4] Also see *Critique of Pure Reason*, A58 = B196f; A237 = B296; A820 = B848.

[5] In this I have benefitted from Tom Nenon's *Objektivität und endliche Erkenntinis. Kants transzendental-philosophische Erkenntnistheorie der Wahrheit*. Freiburg, 1986.

[6] *Kant Werke*, Akademie Ausgabe, Vol XXIV, p. 81.

[7] Ibid., vol. XXIV/1, p. 387.

[8] Ibid., vol. XXIV/1, p. 388.

[9] Ibid., vol. XXIV/1, p. 389.

[10] Ibid., vol. XXIV/1, p. 392.

[11] Ibid., vol. X, p. 131.

[12] Ibid., vol. XIV, p. 244.

[13] Ibid., vol. XIV, p. 246.

[14] Ibid., vol. XVI, p. 254.

[15] Ibid., vol. XVI, p. 255.

[16] Ibid., vol. XXIV/2, p. 525.

[17] Ibid., vol. XXIV/2, p. 525-6.

[18] K. Cramer, "'Gegeben' und 'Gemacht'. Vorüberlegungen zur Function des Begriffs 'Handlung' in Kants Theorie der Erkenntinis von Objekten," in: Prauss (ed.) *Handlungstheorie und Transzendental-philosophie.* Frankfurt-am-Main: V. Klostermann, 1986, pp. 41-81, esp. p. 45.

[19] For the difference between Kant's version of coherence theory and the more standard versions, see Nenon, *loc. cit*, pp.

[20] *Kant Werke*, vol. XVI, p. 252.

[21] Ibid., p. 259.

[22] Ibid., p. 261.

[23] Ibid., p. 255.

[24] R. Stuhlmann-Laeisz, *Kants Logik*, Berlin, 1976.

[25] *Kant Werke*, vol. XXIV, p. 629.

[26] Stuhlmann-Laeisz, p. 64.

[27] Cp. Thomas Scheffer, *Kants Kriterium der Wahrheit*, Berlin: de Gruyter, 1993, esp. pp. 44-5.

[28] Cp. A201 = B246.

[29] *Kant Werke*, vol. XXI, pp. 90-1.

[30] Ibid., p. 134.

[31] Cp. *Logik Blomberg, Kant Werke*, vol. XXIV/1, §§92 - 93; §100; *Logik Pölitz,* Ibid., vol. XXIV/2, pp. 510, 525.

[32] For the idea of the Kantian object, cp. D. Henrich, *Identität und Objektivität. Eine Untersuchung über Kants transzendentale Deduktion.* Heidelberg, 1976.

[33] Cp. *Critique of Pure Reason*, A328 = B384.

[34] *Kant Werke*, vol. XXI, pp. 90-134 ("die mögliche Erfahrung nur Eine ist")

[35] *Critique of Pure Reason,* A158 = B197.

[36] Cp. *Kant Werke*, vol. XXI, p. 53: "Erfahrung ist asymptotische (Kant says on p. 90 "wie in der Hyperbel") Annäherung zur empirischen Vollständigkeit der Wahrnehmungen" ("welche sie ausmachen, is nie Gewissheit", p. 61 — J.N.M.).

[37] Cp. *Critique of Pure Reason*, "The Postulates of Empirical Thought," esp. A219 = B266, A231 = B284.

13. HEGEL'S CONCEPTS OF NECESSITY

Permit me to begin with an autobiographical note. Early in my philosophical studies in Göttingen, Nicolai Hartmann brought home to me the importance of studying the modalities, and also the need for unraveling the many different senses in which the modal categories are used without adequately keeping them apart. In his modal theory, Hartmann returned as much to Stoic logic as to Hegel.[1] Much later in life, when I began to read and teach Hegel's *Phenomenology*, I was struck by the fact that while Hegel intends the series of "shapes" of spirit he portrays to be a necessary sequence, the precise sense of that necessity remains unthematised. I felt certain that the necessity of the sequence could not be logical, for other sequences, slight deviations in that sequence, for example, were quite possible and easily conceivable. And yet I was willing to concede that there must be some sense of 'necessity' in which Hegel was right. The third remark is this: one hears *ad nauseam* that, on Hegel's view, world and history are determined by the "bloodless categories" of his logic, that in this necessary framework there is no room for freedom, for the accidental, the contingent and the unpredictable (the only freedom that is possible being recognition of necessity). And yet I felt certain that this could not be Hegel's view — not only in the *Phenomenology* but also in the *Logic*. (Much later, this conviction was confirmed by Dieter Henrich's essay on the place of *Zufall* in Hegel's thought[2], in which Henrich shows not only that Hegel's system makes room for *Zufall*, but that only Hegel's system can do that.) In the following I make an attempt arising from these concerns to think through Hegel's concept of necessity.

Taken as a modality (although Hegel often criticised the idea of modal necessity), a concept of necessity must be internally connected to the other modalities: possibility and actuality, so that one can have a unified theory of them all. Now there appear to be three main traditions of thinking about the modalities: the Aristotelian, the Leibnizian and the Kantian. On the Aristotelian theory, possibility (as *potentia*) and actuality are stages in the development of a thing. The oak is potentially there in the acorn, it becomes actual. One and the same thing is at first possible, then becomes actual. Note that in this account necessity does not have a clearly defined place. On the Leibnizian theory, more is possible than is actual, and more is actual than is necessary. Possibilities exist first (in God's mind) before they are actualized, so far they are stages (as in the Aristotelian theory), but not of a thing's development. The necessary is what is true in all possible worlds. The Kantian theory — the least ontological of the three — rejects unactualised possibilities as well as the idea of mere potentialities to be actualised. Extensionally, the modalities coincide. One and the same thing is possible, actual and also necessary — possible, when it agrees with the forms of understanding, actual when it agrees with the matter, and necessary when it agrees with both.

Hegel's understanding of the modalities differs from each of these three. But if it is close to any, it is to the Aristotelian. In the other two, there is either a distinction between subject and object (the Kantian modalities are not constitutive of the objects in the full sense, i.e., the sense in which the other categories are), or a theory of unactualised possibilities which also Hegel looked down upon.

With this brief introductory remark, let me turn to how the concept of necessity figures in various parts of Hegel's philosophy.

I. THE ALLEGED NECESSITY IN MATHEMATICS AND THE NATURAL SCIENCES

Kant recognised two kinds of necessary truths: the analytically necessary truths and the synthetic *a priori* truths. As is well known, the latter belong to pure mathematics and pure physics, while the former belong to pure, formal logic. Hegel recognised the momentous importance of Kant's discovery of *a priori* synthesis, but he did not locate it, as Kant did, in either mathematics or in pure physics. To begin with, it is therefore appropriate to briefly recall Hegel's view of mathematics and physics.

Hegel on Mathematics: As against Kant, Hegel maintains that there is no necessity in mathematical knowledge, inasmuch as (a) the process of mathematical proof is external to the mathematical object, the proof being only a subjective process, "an affair of knowledge", "an external function so far as the subject matter dealt with is concerned"[3] ; (b) the construction — again contrary to Kant — is not necessary (in Geometry) and is "imposed on the subject matter"; (c) since mathematics is concerned with quantity, the relationships amongst mathematical quantities are philosophically inessential and superficial, and since "knowledge advances along the lines of bare equality, or abstract identity"[4], mathematics is purely formal; and (d) even in its applied form, mathematics is not able to take time seriously. Arithmetic is analytic[5], proof is only going through an operation. Proving that 7 + 5 = 12 is only following the rules for constructing and adding numbers. There is, in this proof, no transition to an other, what we have is mere repetition of the same operation, an operation which, Hegel says, is so "*gedankenlos*" that a machine can do it.[6] Just because geometrical constructions are external to the subject matter, they are synthetic — but not synthetic *a priori*. They lack necessity. The same fact — namely, that geometry is based on intuition — which Kant advances in favor of the thesis that geometry is synthetic and necessary (since the intuition of space itself is *a priori*), Hegel uses against geometry as a science, for "durch das Anschauen keine Wissenschaft zustande komme, sondern allein durch Denken".[7]

Necessity in Physics: Again, as contrasted with Kant's philosophy of science, Hegel rejects the thesis that there is genuine necessity in the laws of nature. First of all, as he argues in the *Phenomenology*, the plurality of scientific laws, as long as they remain a plurality, are contingently related to each other. To satisfy understanding, they must coalesce into one Law, in which case they would gradually lose their specific contents and the one Law would become more and more abstract — in the long run being nothing more than the bare concept of the Law.[8] Thus there are two possibilities, he tells us[9]: on the one hand, a law must consist of a relation between two or more independent contents; on the other, with the disappearance of these differences, we shall have "a simple withdrawal into itself" of the Law; the bare force, the bare attraction, or simple electricity, is all that we shall have. That positive electricity implies negative electricity is only analytically necessary. There is however no necessity why electricity should differentiate itself into the positive and the negative. Hegel's point at this place is that scientific laws move irremediably between a formulation which preserves necessity while rendering that necessity analytic, and a formulation in which independent entities are brought into a contingent relation, and so which makes the laws synthetic and empirical. Thus in the laws of motion, time and space, distance and velocity — entities which are independent of, and indifferent to one another — are brought to a relation that is not necessary, in which case "motion is . . . no more than their superficial relation." The necessity, then, if at all we can speak of it here, is "merely verbal, is thus a rehearsal of the moments which make up the cycle of necessity."[10] There is no necessary differentiation into the component contents in the nature of the object itself. A similar point is made by Hegel in the chapter on "Observation of organic nature" in the *Phenomenology*: when claims are made that air, water, earth and climate determine features of organic life, Hegel tells us that we do not have more than empirical correlation, we do not find any inner necessity amongst the contents, one can save necessity, as before, only at the cost of rendering it analytic and trivial.[11]

What Hegel is clearly looking for is a concept of necessity which holds together real differences in a unity — a unity which is such that the differences are its internal differentiations. Such a necessity would clearly be synthetic. He does not find it in the natural sciences. In the absence of such synthetic necessity the philosopher of science can only move between regarding scientific laws as empirical, contingent generalisations, and making them, by suitable definitions, analytically necessary.

II. FORMAL LOGIC

It seems from the foregoing that there is at least one concept of necessity which Hegel accepts, however unsatisfying and impoverished that may be: this is the concept of formal-logical and analytical necessity. Hegel's critique of formal logic may be, for my present purpose, represented in two stages: first, as a critique of formal logic as understood and practised by the formal logicians; secondly, as a critique of formal logic insofar as formal logic is taken up into Hegel's *Science of Logic* as one of its parts, namely, the Logic of *Begriff*. The critique of formal logic is also a critique of its claim to be scientific knowledge. Apart from the fact that formal logic deals with abstract forms to which contents are external, and cannot put them back together ("What is purely formal without reality is *ens intellectus*, or empty abstraction"[12]), Hegel insists that formal logic simply finds the forms as *given*, as what merely *are*, "a multitude of detached necessities", which do not constitute a true system in which they are developed by pure thought from within itself, i.e., they are advanced without rigorous and systematic derivation. Since formal logic is not concerned with truth, it can at best have the value of a "natural-historical description of appearances of thought, as they are to be found"[13]. However, in his *Logic*, Hegel did try to take up the forms of formal logic and transform them into the categories of his own logic. A good example of this is found in the Logic of *Begriff*: but here, as with every other category, he finds incoherencies and contradictions combined with something of positive value, which makes them fluid, developing beyond them into the category of objectivity, and eventually to the *Idea*.

Necessity is taken to characterise not only the so-called logical truths, but also the so-called logical principles or "laws of thought". Necessity also appears within logic as characterising certain forms of judgment. It would be worthwhile to take a quick look at them. While analytic propositions are necessary in the sense of being tautologous, Hegel does not regard the so-called laws of thought to be analytic. The principle of Identity itself is, on his view, synthetic,[14] for it contains within itself the movement of reflection by which the identity shows itself as the disappearance of otherness. The same holds good of the law of contradiction: it contains, in its very expression, not alone the empty and simple self-sameness, not also the mere other in general, but the absolute inequality or contradiction. Each of the two laws as a matter of fact contains within itself more than what is intended, i.e., absolute difference. Thus the formal logical principles conceal within their texture movements of reflection, and so their synthetic necessity — other than the merely tautological character which alone formal logic is able to see — lies in these movements of reflection.

As far as the forms of judgment are concerned, Hegel recognises a class of judgments called by him judgments of necessity. Every judgment, according to Hegel, has this form: the individual is the universal (the subject is the predicate). An identity between the two components is asserted.[15] By saying "This rose is red", if Hegel is right, we are not saying that we attach redness to the rose, but rather that the property attaches to the object. Although formal logic makes it appear as if the judgment is contingent, really it is necessary.[16] The judgment of necessity is one in which the predicate represents, to begin with, the essential character of the subject (as in "Gold is a metal"), which is the so-called categorical judgment (the unmediated judgment of necessity).

But this judgment is deficient (with respect to its necessity) inasmuch as not only gold but also copper, silver and iron are also metals. In the hypothetical judgment (If A, then B), the necessity is between the cause and its effect, or between ground and consequent: the actuality of one is not its own, but the being of the other.[17] What the judgment expresses is the necessary connection of immediate determinacies (A and B), each of which is only a possibility.[18] The necessity is the inner relation between the two. The necessity is most explicit in the disjunctive judgment (A is either B or C), where the concrete universal or the genus is the subject, which is seen to be the sum total of the species B and C; thus the judgment expresses the unity of the universal and the particular — its content is thus the *Begriff.* Curiously enough, what Hegel calls the judgment of notion is *formally* what is called judgment of modality (assertoric, problematic and apodictic). It is, with respect to its content, regarded by him as judgment in which value predicates are attached to an object (which is one example of Hegel's restricting the form to a certain type of content seemingly arbitrarily). From judgment, Hegel passes on to syllogism. The standard syllogism of the first figure in which an individual (this rose) is connected to a universal (colored) by means of a particular quality (red), is regarded by him to be completely contingent in respect of its content. The middle term can connect the subject with several different universals, and another middle term (blue) can also serve the same purpose as this one. Here, I think, Hegel's point is confused, but I believe what he is after is a conception of the middle which is such that it *alone* can connect the subject with *this* universal (and with none other). Here as elsewhere, his point seems to be that the syllogism, if it satisfies this requirement, becomes tautologous, and if it does not, then it is contingent. We do not thus find in the formal logical forms examples of genuine synthetic *a priori* necessity. It is thus that Hegel characterises the form of syllogism (in formal logic) to be abstract and *begrifflose* form[19], and mathematization of logic to result in a "völlig gehaltleeren und tautologischen Formierung von Sätzen".[20]

III. ESSENTIAL NECESSITY

The same unavoidable oscillation between "essential necessity" on the one hand and causal necessity on the other is also noted by Hegel. The logic of essence posits an essence as the ground of contingent determinations in each finite thing: "das zufällige Sein gehe an sich selbst in seinem Grund zurück", to an absolutely necessary essence[21], which is the sufficient ground of the contingent contents. However, Hegel goes on to show that the relation of ground, regarded formally, has to take one and the same content as the ground and as the grounded, so that necessity is preserved at the cost of being tautological; considered as real ground, the ground and the grounded are different contents, so that the relation becomes external and contingent.[22] (Hegel's examples are: "The ground of a house is its foundation", "Nature is the ground of the world", and "Electricity is the ground of electrical phenomena".[23])

The case with causal necessity is no better. The cause is a cause only in so far as it has an effect; this is the necessity, the actual, i.e., the acting substance: "die Ursache ist daher erst in ihrer Wirkung das wahrhaft Wirkliche und mit sich identische." The effect contains nothing that the cause does not contain, and the cause is nothing that is not its effect. The relation is tautological. But, as soon as — passing over from formal-causal relation to determinate causal relation — the cause and the effect are taken to be determinate contents, the relation becomes contingent.

IV. POSSIBILITY, ACTUALITY, CONTINGENCY AND NECESSITY

A full blown theory of modality is given by Hegel, first, in the Logic of Essence under the heading "The Actuality" (which is preceded by a chapter entitled "The Absolute" which ends with a discussion of "The Modes of the Absolute" suggesting that in Hegel's mind a discussion of the modalities was triggered by a discussion of Spinozistic "modes".) There are, on Hegel's account at this place, three forms of (or stages of thought about the) modalities: formal, real, and the absolute.

A. Formal Modalities

a. Actuality is formal in so far as it is only immediate, unreflected Being. It is opposed to the actuality of a definite content (which is real actuality). It is also opposed to the totality of all actualities (which is absolute actuality). Formal actuality has its being-in-itself in Possibility (also formal), so that what is actual is also possible. Actuality contains possibility, but this 'containing' is not a real containing, but only logical. Thus formal actuality is also possibility (what is actual is possible) and so non-actuality. Possibility is actuality which is reflected onto itself and thus non-actuality. Thus we have a contradiction.

b. Possibility: Detached from actuality, merely formal or logical possibility has no concreteness, it is abstractly posited as being itself. This abstract logical possibility has two aspects: positively, it is itself, an essentiality. In its negative aspect, it is nothing autonomous, but is completed only in actuality. Thus it is only an appearance of actuality. In itself it is the negated actuality. Thus we have two mutually contradictory propositions: Everything is possible (or rather, everything that is consistent is possible), and nothing is possible. The latter follows from the fact that if A is possible, so is non-A. The possibility of non-A is contained in the possibility of A. We have again a contradiction, which is resolved, at the level of the modalities, by the thesis that only the actual is possible.

c. Accidentality: the actuality and possibility have become one. The mere possibility and the mere actuality are annulled. The result is contingency. The contingent has now two aspects: it is the contingent actuality, i.e., an actuality without any ground. But as contingently possible, its possibility has its ground in another. Thus writes Hegel: "The contingent has no ground, just because it is contingent, and equally well it has a ground just because it is contingent." Contingency is "this absolute unrest of becoming of these two determinations." This unrest finds its solution in formal necessity. The ground lies within itself.

d. Necessity: The formally necessary has its ground in itself. It is both groundless and grounded simultaneously — not unlike the contingent. Only the necessary has its ground within itself.

But we still do not understand how the necessary, i.e., the logically and formally necessary is also actual and possible *simultaneously*. How can it "contain" within itself the other two modalities? This leads us to the real modalities.

B. The Real Modalities:

a. Actuality: Real actuality is determinate actuality. Formal actuality was mere being. Real actuality acts (*das Wirkliche wirkt*), and in so far as it works, it works on an other. It is the thing with its many properties, the existing world. As actuality, it is simultaneously the being-in-itself and reflection on to itself. It contains its own possibility, but the possibility is not formal-logical but real.

b. Possibility: Real possibility is not mere logical consistency, but the totality of conditions which first make a thing possible. In this sense, only the actual is possible. We do not have any longer a "limitless plurality" of formal possibilities ("everything is possible"), but only what is actual is really possible. If the formal possibility of A implies the formal possibility of non-A, the real possibility of A implies that the real impossibility of non-A and the real actuality of A. What this means is that if the totality of conditions which give rise to A is present, then A becomes actual and non-A becomes impossible.

c. Necessity: Thus what is really possible is also really necessary. With the totality of conditions present, it could not be otherwise. (In this sense, Hegel can say that real necessity is the identity of real possibility and real actuality.) Under these very circumstances, nothing else could follow. Real possibility and real actuality differ only apparently. At the same time, this necessity is only relative, it presupposes that from which it takes its start, namely, the contingent. Its content is limited. It is not yet the totality. The conditions which bring about the actuality are themselves contingent, i.e., they do not have their grounds in themselves. They all originate in something contingent. Necessary consequences follow from contingent conditions.

C. Absolute Necessity

This contradiction leads to absolute necessity. Real necessity is in itself contingency. As immediate, given entities, the conditions which give rise to a real entity constitute the real necessity which is itself actuality, though also necessary actuality. In absolute necessity, the opposition between contingency and necessity is annulled. The absolute necessity is also absolute contingency. It is groundless — it is simply because it is. It is its own ground.[24]

We need not tarry to further interpret this chapter from the *Logic*. It may be construed as Hegel's commentary on the traditional cosmological proof for God's existence[25]: the contingent needs absolute necessity. But if the absolute necessity is the condition for the contingent, then it also becomes relational. So one may want to understand the absolute necessity as Spinoza's substance. It is also possible, in order to avoid an acosmic world, to interpret absolute necessity as a Leibnizian monad. Be that as it may, we have learned several things about Hegel's modal doctrines:

1. Hegel's philosophy is opposed to the contemporary glorification of possibility best expressed in Heidegger's statement to the effect that possibility is higher than actuality.[26] In Hegel's view, mere possibility is a deficient mode.

2. In agreement with Kant, Hegel holds that it is the actual which is possible and also necessary (and also contingent). In other words, the actual and the necessary are not extensionally narrower subsets of the possible, rather they are the same entity.

3. How then are we to understand the locution that the actual *contains* possibility and that the necessary *contains* both possibility and actuality? The Kantian deflation of the modalities into a subjective status as contrasted with the other constitutive categories won't do. They are also as objective as the others. However, the locution of "containing" and of "also" (in "The same thing is actual and *also* possible and *also* necessary") shows that like all categories, the modal categories also are vitiated by contradictions and have to be overcome.

4. As Nicolai Hartmann most clearly pointed out, the talk of "possibility", "actuality", "contingency" and "necessity" is equivocal, and means different things for different domains and for different levels of discourse.

5. The ideas of necessity and contingency are inseparably connected, such that all necessity refers back to a contingency and all contingency to a necessity constituted by the totality of conditions. You cannot have then, in Hegelian thinking, a necessary system which has no place for contingency. As Dieter Henrich perhaps most clearly said, it is just the reverse: only Hegel's system has necessarily a place for the contingent.

6. Finally, we need to remember that even absolute necessity is not the final concept of the Absolute in Hegel's *Logic*. Belonging as it does to the Logic of Essence, it is annulled and *aufgehoben* in the Logic of *Begriff*, and is taken over into the idea of freedom. It is not enough to say, as is often done — following Marx — that freedom is the recognition of necessity, for one needs to ask, "recognition by whom?" It is more correct to say that freedom is the truth of necessity.[27] In this truth, both abstract or real necessity and absolute necessity are transcended. The ultimate reality, in the system, is not a series which unfolds itself in an inexorably necessary sequence, it is rather a freedom which uses both necessities (of various orders) and contingencies to actualise itself, and for the entire system the best one can say is: it is such and such because of itself, groundless and free.[28]

V

Recalling Eugen Fink's distinction between thematic and operative concepts of a philosophy, we can say that up until now we have looked at the concepts of necessity which Hegel had thematised. But he also operates with concepts of necessity without thematising them, i.e., without explicitly focusing upon those concepts. There are at least three places where such operative concepts of necessity show themselves: in the *Phenomenology*, in the *Logic* and in the Philosophy of History.

1. In the *Phenomenology*, Hegel says that the completeness of the forms of appearing consciousness will result from the necessity of the progression and interconnection itself.[29] Elsewhere he says, reason demands its satisfaction with respect to form: this form is necessity in general.[30] At another place[31], he speaks of the cold, progressing necessity of the *Sache*. When he speaks of the process which "creates its own moments"[32], the moments are: the being-in-

itself of abstract possibility, the being-for-itself of concrete actuality, and the being-in-and-for-itself of absolute necessity. One can say that Life is the substance as possibility and as positive, self-consciousness is substance as actuality and as negative, and duplication of self-consciousness as the unity of both in substance as Necessity, i.e., as Spirit. Leaving out for the present this last concept of necessity and its relation to the 'absolute necessity' of the system, my concern is the sort of necessity that is taken to characterise the sequence of the forms of the spirit. It is surely not logical necessity, not the real necessity (of the *Logic*) for obvious reasons. Are not other sequences possible? Could it be the merely contingent fact — if it is a fact — that human consciousness has developed in history in that order and through those stages? Could the necessity claimed on its behalf lie in the possibility of reflectively demonstrating *how* the preceding shapes lead to the succeeding ones? In which case, philosophy would be able to show the inner necessity as a reflective necessity, whereas actual history itself is a series of contingencies? All depends upon the idea of a "necessary transition". What sort of necessity is this last mentioned one?

2. A parallel question arises with regard to *Logic*, where the categories are said to unfold in a necessary manner. What could this "necessary" mean, especially when necessity itself is a category? To call it dialectical necessity throws some light on Hegel's procedure, but it does not fully answer the question. The sense of 'necessity' in which from out of a category A, another non-A emerges, and out of the contradiction a third one, B, necessarily comes on the scene is precisely what needs to be clarified without falling back either on the concept of necessity available in formal logic or on the concepts of necessity that are available *within* Hegel's own logic as categories. Perhaps the only category available for understanding the totality of the system is "Freedom" — the idea of a free self-development of the *Begriff*. The idea of necessity as applied to this process is unavoidably misleading, excepting insofar as freedom and necessity entail each other, in Hegel's system.

3. It is in the light of this that we have to understand the nature of Hegel's thesis about history. History is full of contingent truths (on Hegel's own admission), but it is in and through these contingencies that a higher purpose — i.e., consciousness of freedom — is being realised. In the process of this realisation, there is no such *necessity* as will enable us to predict the future (excepting the trivial consequence that the ultimate goal is consciousness of freedom), unlike in Marxism. To say that the Absolute Spirit seeks to know itself in and through history, is neither to weld all history into a grand

monolithic process nor to construe the march of spirit as the irrepressible and uncontrollable march of the Car of Juggernaut. For spirit, in the first place, is freedom (and not an inert force), and spirit works in multifarious ways, through leaps, breaks and ruptures. To assert the Hegelian thesis is not to describe a real process at the level of describing reality, but to comprehend the real history under a large philosophical category, which, as a concrete universal, is inseparable from contingent particularities.

Thinking about the idea of a concrete universal helps us to comprehend the relation of the *Begriff* to particular contingencies. If a concrete universal is one for which every difference is essential (as opposed to the abstract universal which excludes all differences), and if the *Begriff* as Absolute Idea is the only perfect concrete universal, it follows that for the Absolute Idea all contingencies are essential. This last should not be misconstrued to suggest that for the Absolute Idea there are only essential features (the seemingly contingent particularities also being essential). It should mean, if we are to preserve the concreteness of the universal, that it is essential for the life of the spirit that there be contingent particularities, and therefore, that 'necessity' could not be the final truth of things, although being one of the subordinate categories, necessity would belong as a component of the way the spirit works.

NOTES

[1] Nicolai Hartmann, *Möglichkeit und Wirklichkeit*, Meisenheim am Glan: Anton Hain, 2nd edn., 1949.

[2] Dieter Henrich, "Hegels Theorie über den Zufall" in: *Hegel im Kontext*, Frankfurt am Main: Suhrkemp Verlag, 1971. Also cp. Ph. of M., p. 100: "As regards truths...no necessity."

[3] Hegel, *The Phenomenology of Mind*, E. tr. J.B. Baillie. London: George Alten & Unwin. 1966. p.101.

[4] *Phenomenology of Mind*, p.103.

[5] *Wissenschaft d. Logik* II, herausg. G. Lasson, Leipzig: Felix Meirer, 1934, 445.

[6] Ibid, 447.

[7] Ibid, 472.

[8] *Phenomenology of Mind*, p.196.

[9] Ibid, pp.197-198.

[10] Ibid, p.200.

[11] Ibid, pp.294-295

[12] Ibid, p.329.

[13] *Wissenschaft d. Logik* II, 234.

[14] Ibid, 32.

[15] *Encyclopädie der philosophischen Wissenschaften*, herausg. J. Hoffmeister, Leipzig: Felix Meirer, 5th edn., 1949, § 166.

[16] Ibid., § 177.

[17] Ibid, § 177.

[18] *Wissenschaft d. Logik* II, 296.

[19] Ibid, 330.

[20] Ibid, 333.

[21] Ibid, 62.

[22] Ibid, 84.

[23] *Encyc.*, § 121.

[24] *Wissenschaft d. Logik* II, 215.

[25] Schmidt, Gerhardt, "Das Spiel der Modalitäten und die Macht der Notwendigkeit", in Irving Fetscher, ed., *Hegel in der Sicht der neueren Forschung*, Darmstadt: Wiss. Buchgesellschaft, 1973, p.199.

[26] Heidegger, M., *Sein und Zeit*, Tübingen: Max Niemeyer, 7th edn., 1953, p.38.

[27] *Encyc*, § 158.

[28] Also see *Encyc*, § 35 Zusatz.

[29] *Phenomenology of Mind*, p.137.

[30] Ibid, p.76.

[31] Ibid, p.72.

[32] Ibid, p.105.

INDEX

S

T

W

SYNTHESE LIBRARY

139. L. Nowak, *The Structure of Idealization.* Towards a Systematic Interpretation of the Marxian Idea of Science. 1980 ISBN 90-277-1014-7
140. C. Perelman, *The New Rhetoric and the Humanities.* Essays on Rhetoric and Its Applications. Translated from French and German. With an Introduction by H. Zyskind. 1979 ISBN 90-277-1018-X; Pb 90-277-1019-8
141. W. Rabinowicz, *Universalizability.* A Study in Morals and Metaphysics. 1979 ISBN 90-277-1020-2
142. C. Perelman, *Justice, Law and Argument.* Essays on Moral and Legal Reasoning. Translated from French and German. With an Introduction by H.J. Berman. 1980 ISBN 90-277-1089-9; Pb 90-277-1090-2
143. S. Kanger and S. Öhman (eds.), *Philosophy and Grammar.* Papers on the Occasion of the Quincentennial of Uppsala University. 1981 ISBN 90-277-1091-0
144. T. Pawlowski, *Concept Formation in the Humanities and the Social Sciences.* 1980 ISBN 90-277-1096-1
145. J. Hintikka, D. Gruender and E. Agazzi (eds.), *Theory Change, Ancient Axiomatics and Galileo's Methodology.* Proceedings of the 1978 Pisa Conference on the History and Philosophy of Science, Volume I. 1981 ISBN 90-277-1126-7
146. J. Hintikka, D. Gruender and E. Agazzi (eds.), *Probabilistic Thinking, Thermodynamics, and the Interaction of the History and Philosophy of Science.* Proceedings of the 1978 Pisa Conference on the History and Philosophy of Science, Volume II. 1981 ISBN 90-277-1127-5
147. U. Mönnich (ed.), *Aspects of Philosophical Logic.* Some Logical Forays into Central Notions of Linguistics and Philosophy. 1981 ISBN 90-277-1201-8
148. D. M. Gabbay, *Semantical Investigations in Heyting's Intuitionistic Logic.* 1981 ISBN 90-277-1202-6
149. E. Agazzi (ed.), *Modern Logic – A Survey.* Historical, Philosophical, and Mathematical Aspects of Modern Logic and Its Applications. 1981 ISBN 90-277-1137-2
150. A. F. Parker-Rhodes, *The Theory of Indistinguishables.* A Search for Explanatory Principles below the Level of Physics. 1981 ISBN 90-277-1214-X
151. J. C. Pitt, *Pictures, Images, and Conceptual Change.* An Analysis of Wilfrid Sellars' Philosophy of Science. 1981 ISBN 90-277-1276-X; Pb 90-277-1277-8
152. R. Hilpinen (ed.), *New Studies in Deontic Logic.* Norms, Actions, and the Foundations of Ethics. 1981 ISBN 90-277-1278-6; Pb 90-277-1346-4
153. C. Dilworth, *Scientific Progress.* A Study Concerning the Nature of the Relation between Successive Scientific Theories. 3rd rev. ed., 1994 ISBN 0-7923-2487-0; Pb 0-7923-2488-9
154. D. Woodruff Smith and R. McIntyre, *Husserl and Intentionality.* A Study of Mind, Meaning, and Language. 1982 ISBN 90-277-1392-8; Pb 90-277-1730-3
155. R. J. Nelson, *The Logic of Mind.* 2nd. ed., 1989 ISBN 90-277-2819-4; Pb 90-277-2822-4
156. J. F. A. K. van Benthem, *The Logic of Time.* A Model-Theoretic Investigation into the Varieties of Temporal Ontology, and Temporal Discourse. 1983; 2nd ed., 1991 ISBN 0-7923-1081-0
157. R. Swinburne (ed.), *Space, Time and Causality.* 1983 ISBN 90-277-1437-1
158. E. T. Jaynes, *Papers on Probability, Statistics and Statistical Physics.* Ed. by R. D. Rozenkrantz. 1983 ISBN 90-277-1448-7; Pb (1989) 0-7923-0213-3
159. T. Chapman, *Time: A Philosophical Analysis.* 1982 ISBN 90-277-1465-7
160. E. N. Zalta, *Abstract Objects.* An Introduction to Axiomatic Metaphysics. 1983 ISBN 90-277-1474-6
161. S. Harding and M. B. Hintikka (eds.), *Discovering Reality.* Feminist Perspectives on Epistemology, Metaphysics, Methodology, and Philosophy of Science. 1983 ISBN 90-277-1496-7; Pb 90-277-1538-6

SYNTHESE LIBRARY

162. M. A. Stewart (ed.), *Law, Morality and Rights.* 1983 ISBN 90-277-1519-X
163. D. Mayr and G. Süssmann (eds.), *Space, Time, and Mechanics.* Basic Structures of a Physical Theory. 1983 ISBN 90-277-1525-4
164. D. Gabbay and F. Guenthner (eds.), *Handbook of Philosophical Logic.* Vol. I: Elements of Classical Logic. 1983 ISBN 90-277-1542-4
165. D. Gabbay and F. Guenthner (eds.), *Handbook of Philosophical Logic.* Vol. II: Extensions of Classical Logic. 1984 ISBN 90-277-1604-8
166. D. Gabbay and F. Guenthner (eds.), *Handbook of Philosophical Logic.* Vol. III: Alternative to Classical Logic. 1986 ISBN 90-277-1605-6
167. D. Gabbay and F. Guenthner (eds.), *Handbook of Philosophical Logic.* Vol. IV: Topics in the Philosophy of Language. 1989 ISBN 90-277-1606-4
168. A. J. I. Jones, *Communication and Meaning.* An Essay in Applied Modal Logic. 1983 ISBN 90-277-1543-2
169. M. Fitting, *Proof Methods for Modal and Intuitionistic Logics.* 1983 ISBN 90-277-1573-4
170. J. Margolis, *Culture and Cultural Entities.* Toward a New Unity of Science. 1984 ISBN 90-277-1574-2
171. R. Tuomela, *A Theory of Social Action.* 1984 ISBN 90-277-1703-6
172. J. J. E. Gracia, E. Rabossi, E. Villanueva and M. Dascal (eds.), *Philosophical Analysis in Latin America.* 1984 ISBN 90-277-1749-4
173. P. Ziff, *Epistemic Analysis.* A Coherence Theory of Knowledge. 1984 ISBN 90-277-1751-7
174. P. Ziff, *Antiaesthetics.* An Appreciation of the Cow with the Subtile Nose. 1984 ISBN 90-277-1773-7
175. W. Balzer, D. A. Pearce, and H.-J. Schmidt (eds.), *Reduction in Science.* Structure, Examples, Philosophical Problems. 1984 ISBN 90-277-1811-3
176. A. Peczenik, L. Lindahl and B. van Roermund (eds.), *Theory of Legal Science.* Proceedings of the Conference on Legal Theory and Philosophy of Science (Lund, Sweden, December 1983). 1984 ISBN 90-277-1834-2
177. I. Niiniluoto, *Is Science Progressive?* 1984 ISBN 90-277-1835-0
178. B. K. Matilal and J. L. Shaw (eds.), *Analytical Philosophy in Comparative Perspective.* Exploratory Essays in Current Theories and Classical Indian Theories of Meaning and Reference. 1985 ISBN 90-277-1870-9
179. P. Kroes, *Time: Its Structure and Role in Physical Theories.* 1985 ISBN 90-277-1894-6
180. J. H. Fetzer, *Sociobiology and Epistemology.* 1985 ISBN 90-277-2005-3; Pb 90-277-2006-1
181. L. Haaparanta and J. Hintikka (eds.), *Frege Synthesized.* Essays on the Philosophical and Foundational Work of Gottlob Frege. 1986 ISBN 90-277-2126-2
182. M. Detlefsen, *Hilbert's Program.* An Essay on Mathematical Instrumentalism. 1986 ISBN 90-277-2151-3
183. J. L. Golden and J. J. Pilotta (eds.), *Practical Reasoning in Human Affairs.* Studies in Honor of Chaim Perelman. 1986 ISBN 90-277-2255-2
184. H. Zandvoort, *Models of Scientific Development and the Case of Nuclear Magnetic Resonance.* 1986 ISBN 90-277-2351-6
185. I. Niiniluoto, *Truthlikeness.* 1987 ISBN 90-277-2354-0
186. W. Balzer, C. U. Moulines and J. D. Sneed, *An Architectonic for Science.* The Structuralist Program. 1987 ISBN 90-277-2403-2
187. D. Pearce, *Roads to Commensurability.* 1987 ISBN 90-277-2414-8
188. L. M. Vaina (ed.), *Matters of Intelligence.* Conceptual Structures in Cognitive Neuroscience. 1987 ISBN 90-277-2460-1

189. H. Siegel, *Relativism Refuted.* A Critique of Contemporary Epistemological Relativism. 1987 ISBN 90-277-2469-5
190. W. Callebaut and R. Pinxten, *Evolutionary Epistemology.* A Multiparadigm Program, with a Complete Evolutionary Epistemology Bibliograph. 1987 ISBN 90-277-2582-9
191. J. Kmita, *Problems in Historical Epistemology.* 1988 ISBN 90-277-2199-8
192. J. H. Fetzer (ed.), *Probability and Causality.* Essays in Honor of Wesley C. Salmon, with an Annotated Bibliography. 1988 ISBN 90-277-2607-8; Pb 1-5560-8052-2
193. A. Donovan, L. Laudan and R. Laudan (eds.), *Scrutinizing Science.* Empirical Studies of Scientific Change. 1988 ISBN 90-277-2608-6
194. H.R. Otto and J.A. Tuedio (eds.), *Perspectives on Mind.* 1988 ISBN 90-277-2640-X
195. D. Batens and J.P. van Bendegem (eds.), *Theory and Experiment.* Recent Insights and New Perspectives on Their Relation. 1988 ISBN 90-277-2645-0
196. J. Österberg, *Self and Others.* A Study of Ethical Egoism. 1988 ISBN 90-277-2648-5
197. D.H. Helman (ed.), *Analogical Reasoning.* Perspectives of Artificial Intelligence, Cognitive Science, and Philosophy. 1988 ISBN 90-277-2711-2
198. J. Woleński, *Logic and Philosophy in the Lvov-Warsaw School.* 1989 ISBN 90-277-2749-X
199. R. Wójcicki, *Theory of Logical Calculi.* Basic Theory of Consequence Operations. 1988 ISBN 90-277-2785-6
200. J. Hintikka and M.B. Hintikka, *The Logic of Epistemology and the Epistemology of Logic.* Selected Essays. 1989 ISBN 0-7923-0040-8; Pb 0-7923-0041-6
201. E. Agazzi (ed.), *Probability in the Sciences.* 1988 ISBN 90-277-2808-9
202. M. Meyer (ed.), *From Metaphysics to Rhetoric.* 1989 ISBN 90-277-2814-3
203. R.L. Tieszen, *Mathematical Intuition.* Phenomenology and Mathematical Knowledge. 1989 ISBN 0-7923-0131-5
204. A. Melnick, *Space, Time, and Thought in Kant.* 1989 ISBN 0-7923-0135-8
205. D.W. Smith, *The Circle of Acquaintance.* Perception, Consciousness, and Empathy. 1989 ISBN 0-7923-0252-4
206. M.H. Salmon (ed.), *The Philosophy of Logical Mechanism.* Essays in Honor of Arthur W. Burks. With his Responses, and with a Bibliography of Burk's Work. 1990 ISBN 0-7923-0325-3
207. M. Kusch, *Language as Calculus vs. Language as Universal Medium.* A Study in Husserl, Heidegger, and Gadamer. 1989 ISBN 0-7923-0333-4
208. T.C. Meyering, *Historical Roots of Cognitive Science.* The Rise of a Cognitive Theory of Perception from Antiquity to the Nineteenth Century. 1989 ISBN 0-7923-0349-0
209. P. Kosso, *Observability and Observation in Physical Science.* 1989 ISBN 0-7923-0389-X
210. J. Kmita, *Essays on the Theory of Scientific Cognition.* 1990 ISBN 0-7923-0441-1
211. W. Sieg (ed.), *Acting and Reflecting.* The Interdisciplinary Turn in Philosophy. 1990 ISBN 0-7923-0512-4
212. J. Karpiński, *Causality in Sociological Research.* 1990 ISBN 0-7923-0546-9
213. H.A. Lewis (ed.), *Peter Geach: Philosophical Encounters.* 1991 ISBN 0-7923-0823-9
214. M. Ter Hark, *Beyond the Inner and the Outer.* Wittgenstein's Philosophy of Psychology. 1990 ISBN 0-7923-0850-6
215. M. Gosselin, *Nominalism and Contemporary Nominalism.* Ontological and Epistemological Implications of the Work of W.V.O. Quine and of N. Goodman. 1990 ISBN 0-7923-0904-9
216. J.H. Fetzer, D. Shatz and G. Schlesinger (eds.), *Definitions and Definability.* Philosophical Perspectives. 1991 ISBN 0-7923-1046-2
217. E. Agazzi and A. Cordero (eds.), *Philosophy and the Origin and Evolution of the Universe.* 1991 ISBN 0-7923-1322-4

SYNTHESE LIBRARY

218. M. Kusch, *Foucault's Strata and Fields.* An Investigation into Archaeological and Genealogical Science Studies. 1991 ISBN 0-7923-1462-X
219. C.J. Posy, *Kant's Philosophy of Mathematics.* Modern Essays. 1992 ISBN 0-7923-1495-6
220. G. Van de Vijver, *New Perspectives on Cybernetics.* Self-Organization, Autonomy and Connectionism. 1992 ISBN 0-7923-1519-7
221. J.C. Nyíri, *Tradition and Individuality.* Essays. 1992 ISBN 0-7923-1566-9
222. R. Howell, *Kant's Transcendental Deduction.* An Analysis of Main Themes in His Critical Philosophy. 1992 ISBN 0-7923-1571-5
223. A. García de la Sienra, *The Logical Foundations of the Marxian Theory of Value.* 1992 ISBN 0-7923-1778-5
224. D.S. Shwayder, *Statement and Referent.* An Inquiry into the Foundations of Our Conceptual Order. 1992 ISBN 0-7923-1803-X
225. M. Rosen, *Problems of the Hegelian Dialectic.* Dialectic Reconstructed as a Logic of Human Reality. 1993 ISBN 0-7923-2047-6
226. P. Suppes, *Models and Methods in the Philosophy of Science: Selected Essays.* 1993 ISBN 0-7923-2211-8
227. R. M. Dancy (ed.), *Kant and Critique: New Essays in Honor of W. H. Werkmeister.* 1993 ISBN 0-7923-2244-4
228. J. Woleński (ed.), *Philosophical Logic in Poland.* 1993 ISBN 0-7923-2293-2
229. M. De Rijke (ed.), *Diamonds and Defaults.* Studies in Pure and Applied Intensional Logic. 1993 ISBN 0-7923-2342-4
230. B.K. Matilal and A. Chakrabarti (eds.), *Knowing from Words.* Western and Indian Philosophical Analysis of Understanding and Testimony. 1994 ISBN 0-7923-2345-9
231. S.A. Kleiner, *The Logic of Discovery.* A Theory of the Rationality of Scientific Research. 1993 ISBN 0-7923-2371-8
232. R. Festa, *Optimum Inductive Methods.* A Study in Inductive Probability, Bayesian Statistics, and Verisimilitude. 1993 ISBN 0-7923-2460-9
233. P. Humphreys (ed.), *Patrick Suppes: Scientific Philosopher.* Vol. 1: Probability and Probabilistic Causality. 1994 ISBN 0-7923-2552-4
234. P. Humphreys (ed.), *Patrick Suppes: Scientific Philosopher.* Vol. 2: Philosophy of Physics, Theory Structure, and Measurement Theory. 1994 ISBN 0-7923-2553-2
235. P. Humphreys (ed.), *Patrick Suppes: Scientific Philosopher.* Vol. 3: Language, Logic, and Psychology. 1994 ISBN 0-7923-2862-0
     Set ISBN (Vols 233–235) 0-7923-2554-0
236. D. Prawitz and D. Westerståhl (eds.), *Logic and Philosophy of Science in Uppsala.* Papers from the 9th International Congress of Logic, Methodology, and Philosophy of Science. 1994 ISBN 0-7923-2702-0
237. L. Haaparanta (ed.), *Mind, Meaning and Mathematics.* Essays on the Philosophical Views of Husserl and Frege. 1994 ISBN 0-7923-2703-9
238. J. Hintikka (ed.), *Aspects of Metaphor.* 1994 ISBN 0-7923-2786-1
239. B. McGuinness and G. Oliveri (eds.), *The Philosophy of Michael Dummett.* With Replies from Michael Dummett. 1994 ISBN 0-7923-2804-3
240. D. Jamieson (ed.), *Language, Mind, and Art.* Essays in Appreciation and Analysis, In Honor of Paul Ziff. 1994 ISBN 0-7923-2810-8
241. G. Preyer, F. Siebelt and A. Ulfig (eds.), *Language, Mind and Epistemology.* On Donald Davidson's Philosophy. 1994 ISBN 0-7923-2811-6
242. P. Ehrlich (ed.), *Real Numbers, Generalizations of the Reals, and Theories of Continua.* 1994 ISBN 0-7923-2689-X

243. G. Debrock and M. Hulswit (eds.), *Living Doubt*. Essays concerning the epistemology of Charles Sanders Peirce. 1994 ISBN 0-7923-2898-1
244. J. Srzednicki, *To Know or Not to Know*. Beyond Realism and Anti-Realism. 1994 ISBN 0-7923-2909-0
245. R. Egidi (ed.), *Wittgenstein: Mind and Language*. 1995 ISBN 0-7923-3171-0
246. A. Hyslop, *Other Minds*. 1995 ISBN 0-7923-3245-8
247. L. Pólos and M. Masuch (eds.), *Applied Logic: How, What and Why*. Logical Approaches to Natural Language. 1995 ISBN 0-7923-3432-9
248. M. Krynicki, M. Mostowski and L.M. Szczerba (eds.), *Quantifiers: Logics, Models and Computation*. Volume One: Surveys. 1995 ISBN 0-7923-3448-5
249. M. Krynicki, M. Mostowski and L.M. Szczerba (eds.), *Quantifiers: Logics, Models and Computation*. Volume Two: Contributions. 1995 ISBN 0-7923-3449-3
Set ISBN (Vols 248 + 249) 0-7923-3450-7
250. R.A. Watson, *Representational Ideas from Plato to Patricia Churchland*. 1995 ISBN 0-7923-3453-1
251. J. Hintikka (ed.), *From Dedekind to Gödel*. Essays on the Development of the Foundations of Mathematics. 1995 ISBN 0-7923-3484-1
252. A. Wiśniewski, *The Posing of Questions*. Logical Foundations of Erotetic Inferences. 1995 ISBN 0-7923-3637-2
253. J. Peregrin, *Doing Worlds with Words*. Formal Semantics without Formal Metaphysics. 1995 ISBN 0-7923-3742-5
254. I.A. Kieseppä, *Truthlikeness for Multidimensional, Quantitative Cognitive Problems*. 1996 ISBN 0-7923-4005-1
255. P. Hugly and C. Sayward: *Intensionality and Truth*. An Essay on the Philosophy of A.N. Prior. 1996 ISBN 0-7923-4119-8
256. L. Hankinson Nelson and J. Nelson (eds.): *Feminism, Science, and the Philosophy of Science*. 1997 ISBN 0-7923-4162-7
257. P.I. Bystrov and V.N. Sadovsky (eds.): *Philosophical Logic and Logical Philosophy*. Essays in Honour of Vladimir A. Smirnov. 1996 ISBN 0-7923-4270-4
258. Å.E. Andersson and N-E. Sahlin (eds.): *The Complexity of Creativity*. 1996 ISBN 0-7923-4346-8
259. M.L. Dalla Chiara, K. Doets, D. Mundici and J. van Benthem (eds.): *Logic and Scientific Methods*. Volume One of the Tenth International Congress of Logic, Methodology and Philosophy of Science, Florence, August 1995. 1997 ISBN 0-7923-4383-2
260. M.L. Dalla Chiara, K. Doets, D. Mundici and J. van Benthem (eds.): *Structures and Norms in Science*. Volume Two of the Tenth International Congress of Logic, Methodology and Philosophy of Science, Florence, August 1995. 1997 ISBN 0-7923-4384-0
Set ISBN (Vols 259 + 260) 0-7923-4385-9
261. A. Chakrabarti: *Denying Existence*. The Logic, Epistemology and Pragmatics of Negative Existentials and Fictional Discourse. 1997 ISBN 0-7923-4388-3
262. A. Biletzki: *Talking Wolves*. Thomas Hobbes on the Language of Politics and the Politics of Language. 1997 ISBN 0-7923-4425-1
263. D. Nute (ed.): *Defeasible Deontic Logic*. 1997 ISBN 0-7923-4630-0
264. U. Meixner: *Axiomatic Formal Ontology*. 1997 ISBN 0-7923-4747-X
265. I. Brinck: *The Indexical 'I'*. The First Person in Thought and Language. 1997 ISBN 0-7923-4741-2
266. G. Hölmström-Hintikka and R. Tuomela (eds.): *Contemporary Action Theory*. Volume 1: Individual Action. 1997 ISBN 0-7923-4753-6; Set: 0-7923-4754-4

SYNTHESE LIBRARY

267. G. Hölmström-Hintikka and R. Tuomela (eds.): *Contemporary Action Theory.* Volume 2: Social Action. 1997 ISBN 0-7923-4752-8; Set: 0-7923-4754-4
268. B.-C. Park: *Phenomenological Aspects of Wittgenstein's Philosophy.* 1998 ISBN 0-7923-4813-3
269. J. Paśniczek: *The Logic of Intentional Objects.* A Meinongian Version of Classical Logic. 1998 ISBN 0-7923-4880-X
270. P.W. Humphreys and J.H. Fetzer (eds.): *The New Theory of Reference.* Kripke, Marcus, and Its Origins. 1998 ISBN 0-7923-4898-2
271. K. Szaniawski, A. Chmielewski and J. Woleński (eds.): *On Science, Inference, Information and Decision Making.* Selected Essays in the Philosophy of Science. 1998 ISBN 0-7923-4922-9
272. G.H. von Wright: *In the Shadow of Descartes.* Essays in the Philosophy of Mind. 1998 ISBN 0-7923-4992-X
273. K. Kijania-Placek and J. Woleński (eds.): *The Lvov–Warsaw School and Contemporary Philosophy.* 1998 ISBN 0-7923-5105-3
274. D. Dedrick: *Naming the Rainbow.* Colour Language, Colour Science, and Culture. 1998 ISBN 0-7923-5239-4
275. L. Albertazzi (ed.): *Shapes of Forms.* From Gestalt Psychology and Phenomenology to Ontology and Mathematics. 1999 ISBN 0-7923-5246-7
276. P. Fletcher: *Truth, Proof and Infinity.* A Theory of Constructions and Constructive Reasoning. 1998 ISBN 0-7923-5262-9
277. M. Fitting and R.L. Mendelsohn (eds.): *First-Order Modal Logic.* 1998 ISBN 0-7923-5334-X
278. J.N. Mohanty: *Logic, Truth and Modalities.* From a Phenomenological Perspective. 1999 ISBN 0-7923-5550-4

Previous volumes are still available.

KLUWER ACADEMIC PUBLISHERS – DORDRECHT / BOSTON / LONDON